D1176122

Nature Library

MAMMALS

Nature Library

MAMMALS

Robert Burton

Viscount Books

Artists

John Beswick, Michael Clarke, Design Practitioners Ltd., Don Forrest, Peter Morter, David Nockels, Richard Orr, A. Oxenham, Pat Oxenham, David Pratt, John Rignall, George Thompson, Peter Thornley, Tudor Art Agency Ltd., Peter Warner

This edition published by Viscount Books 1984.
Prepared by Newnes Books,
a division of The Hamlyn Publishing Group Limited
84-88 The Centre, Feltham, Middlesex, England,
and distributed for them by
The Hamlyn Publishing Group Limited,
Rushden, Northants, England.

ISBN 0 600 35662 0

Printed in Italy

Photographic Acknowledgements

ANIMAL PHOTOGRAPHY, LONDON: Sally Anne Thompson 71 centre right; BRUCE COLEMAN, UXBRIDGE: 22 right, 36 top, 74 top; Ken Balcomb 66 top; Jen & Des Bartlett 19 top, 20, 46 top right, 48 bottom, 49 centre, 50 top, 51 top left, 56, 60 top, 68 centre; Erwin and Peggy Bauer 30, 37 top left, 40 top; Chris Bonnington 71 bottom; R. & M. Borland 11; Mark Boulton 45 centre, 49 top left; John R. Brownlie 21 top left; Jane Burton 12 top, 12 bottom, 13 bottom, 15 bottom, 28 bottom, 29 bottom, 43 bottom, 44 bottom, 49 top right; Bob and Clara Calhoun 46 centre, 64 bottom; R. I. M. Campbell 47; Eric Crichton 28 top; Gerald Cubitt 46 bottom; Peter Davey 9, 42 bottom, 43 top, 50 bottom, 51 bottom; L. E. Dawson 34; Francisco Erize 21 top right, 24 top, 32 right, 53 top and bottom, 64 top, 65 top, 68 bottom, 69 centre, 75 left; John Fennel 29 top left; Jeff Foott 19 bottom, 41 left, 65 bottom, 69 top; C. B. & D. W. Frith 72 bottom; J. L. G. Grande 37 bottom; S. Halvorsen 27 bottom; Charles Henneghien 46 top left; P. A. Hinchcliffe 76-77; Udo Hirsch 75 right; Jerry L. Hout 24 bottom; Carol Hughes 51 top right; M. P. Kahl 49 bottom; Stephen J. Krasermann 35 top, 73 bottom right; Wayne Lankinen 45 top left; Rocco Longo 71 centre; Lee Lyon 62 right; L. M. Myers 25 bottom; Charlie Ott 26-27, 35 centre and bottom; Dieter and Mary Plage 52 bottom; Andy Purcell 29 top right, 73 bottom left; Hans Reinhard 8, 13 top, 31, 32 left, 33, 38, 42 top, 54 bottom, 70 top and bottom; Leonard Lee Rue III 15 top, 18, 22 left, 37 top right, 40 bottom, 60 bottom, 69 bottom, 72 top right, 73 top; Leonard Lee Rue IV 26 top; W. E. Ruth 26 bottom; James Simon 25 top; Stouffer Productions 27 top; Sullivan and Rogers 21 bottom, 45 bottom; Kunio Takama/Orion Press 62 left; Norman Tomalin 41 right, 63; Simon Trevor 39; Peter Ward 58; Joe van Wormer 54 top, 59 bottom; World Wildlife Fund/Henry Ausloos 45 top right, 52 top; World Wildlife Fund/Al Giddings 66 bottom; World Wildlife Fund/Jacques Gilliéron 6-7, 48 top; World Wildlife Fund/Tim Rautert 74 bottom; G. Ziesler 57, 72 top left; GEOFFREY KINNS 36 centre; JUDY TODD 44 top.

Contents

Introduction

The immense variety of the mammal class is a never-ending source of wonder. Some of these animals hold a fascination because of their appealing physical characteristics, such as the seals, pandas and bears; others attract attention because of their size, such as whales and elephants. There are also those that invoke admiration through their movements; for instances, the acrobatic monkeys, playful dolphins, elegant gazelles and lithesome cats. Other mammals, bats and rats for example, are regarded with distaste for real or imaginary reasons. Yet there are thousands of mammals that are frequently overlooked. They are small, secretive and very often active only at night.

Many of these mammals are being subjected to studies lasting several years, which shows the complexity of their private lives. Mammals are the most intelligent of animals. Their capacity to learn and their ability to adapt to changing environments results in a richness and variety of habits not seen in other species.

The following pages introduce the basic biology of mammals, together with surveys of certain groups and inhabitants of particular habitats.

What is a mammal?

The milk produced by mammals for suckling their young is rich in fats, proteins and milk sugar. It varies in composition with the needs of the young and, seemingly, with the environment. In an arid area, the milk has a high water content, whereas marine and Arctic mammals produce a high fat content. Here the fat, or edible, dormouse is suckling her young.

Opposite top left: the brain of a mammal and a reptile compared. Note the development of the mammal's cerebrum or cerebral hemispheres, the seat of more advanced behaviour such as learning and intelligence.

One unique attribute of mammals, possessed by them and no other animal, is the ability of the mother to produce milk for nourishing the offspring. The young are fed from her mammary glands and the word mammal comes from the Latin *mamma*, meaning breast. Another unique feature is hair. All mammals have some hair – even young whales have a few bristles on the throat.

The scientific word Mammalia was first used by the Swedish naturalist Linnaeus when he classified the Natural World in 1773, and the English version of mammal came into use only in 1826. Even now, 'animal' is often used to describe mammals when it should be used for all kinds of animals from diminutive, single-celled protozoans to giant whales.

There are just over 4000 species of mammals in the world today – that is approximately two-thirds the number of reptile species and about half the total of bird species. Several million years ago there were many more species, but, since then, entire groups of mammals have become extinct. The rate of extinction has speeded up in recent years through the activities of one of the species – man.

Milk production, hair and many improvements in the skeleton, brain and blood and breathing systems have given the mammals advantages over their reptile ancestors. A supply of mother's milk saves a young mammal having to find its own food. It can grow rapidly while under parental protection and it can learn how to look after itself more effectively with parental guidance.

The possession of hair is one of the most significant characteristics of a mammal, because it is related to an essential difference between mammals and reptiles. A covering of hair or fur acts as an insulation to help keep the mammal warm.

A reptile is cold-blooded; but this is misleading. It can warm its body by basking in the sun and try to retain its warmth by burrowing into warm soil at night. It is better to call a reptile 'ectothermic', meaning that it gets its heat from outside its body. Mammals, and birds, are warm-blooded, or more correctly 'endothermic', meaning that their heat comes from inside the body. The heat is generated biochemically, by the oxidation of carbohydrates, and body temperature is regulated within narrow limits (36°C-38°C in most mammals) by a thermostat – the hypothalamus in the brain. The insulation of the fur reduces the amount of energy needed to maintain the high body temperature. Some mammals, however, such as whales,

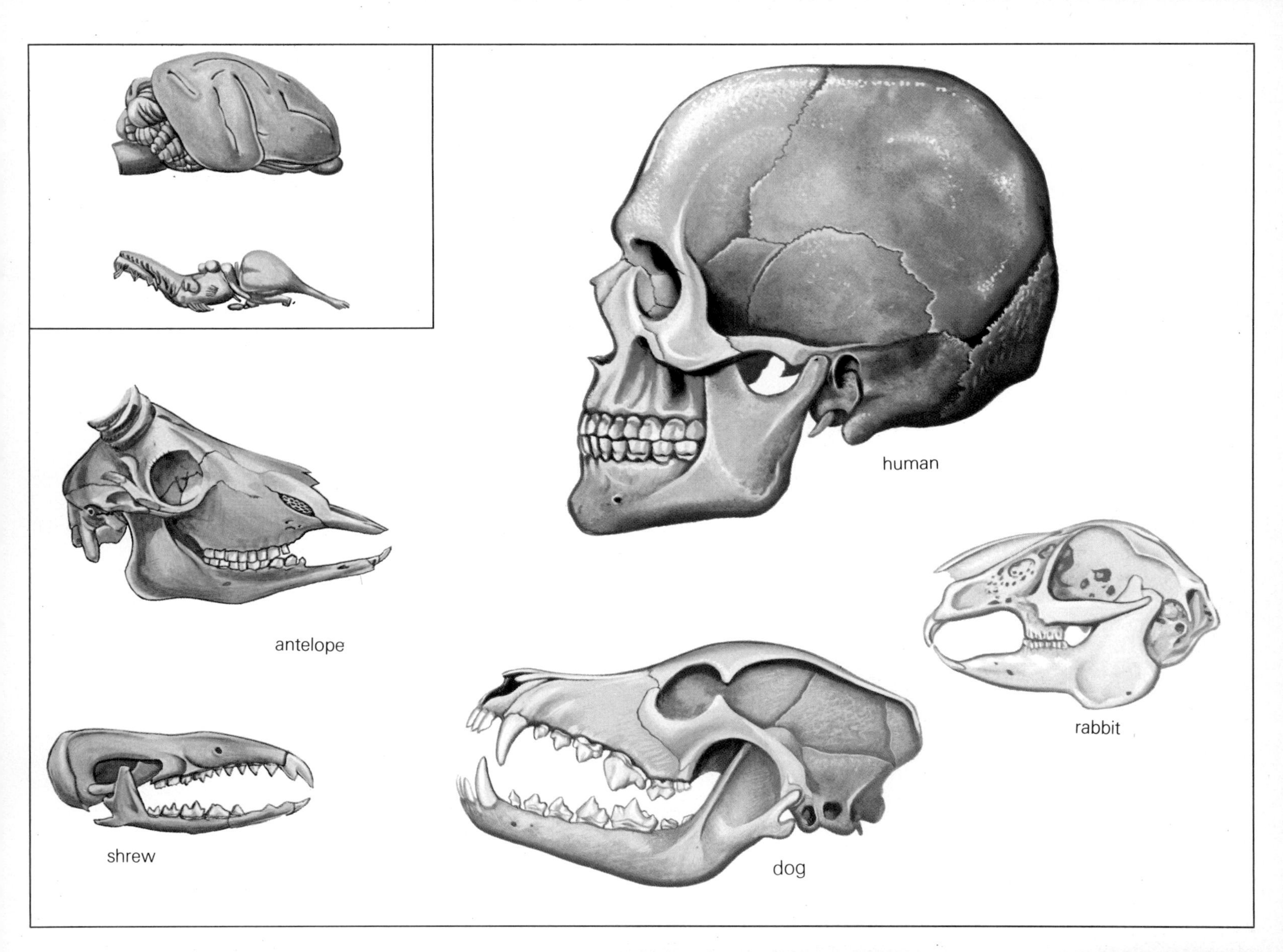

Above: **skulls of mammals showing the variety of teeth. The human is an omnivore with fairly uniform teeth. The vegetarian antelope has flat teeth for crushing leaves and the rabbit also has long incisors for gnawing and nibbling. The shrew has sharp teeth for crunching insects. The dog has stabbing canines, or fangs, and blade-like cheek-teeth for slicing meat.**

Right: **the fangs of this lioness are shown off well when she yawns.**

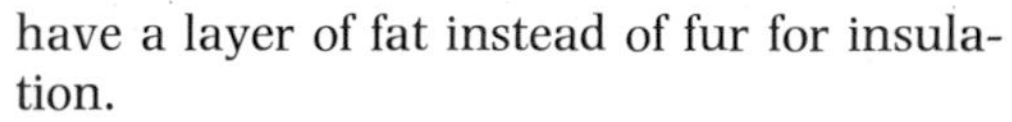

have a layer of fat instead of fur for insulation.

In hot climates and during exercise it may be necessary to cool the body. This is achieved by sweating: sweat glands pour out a watery solution which cools the skin by evaporation. Not all mammals have sweat glands. The desert kangaroo rat, for instance, salivates over its fur for the same purpose, and dogs pant to increase evaporation of saliva from the mouth.

Hair is made of keratin, as are birds' feathers, and must be replaced when worn. At intervals, therefore, old hair or fur is shed and replaced by new hair through the process of moulting, and, in addition, some mammals grow a thicker coat for the winter. There are two basic kinds of hair: soft, thick underfur provides the insulation (sheep's wool is very long underfur) and long, coarse

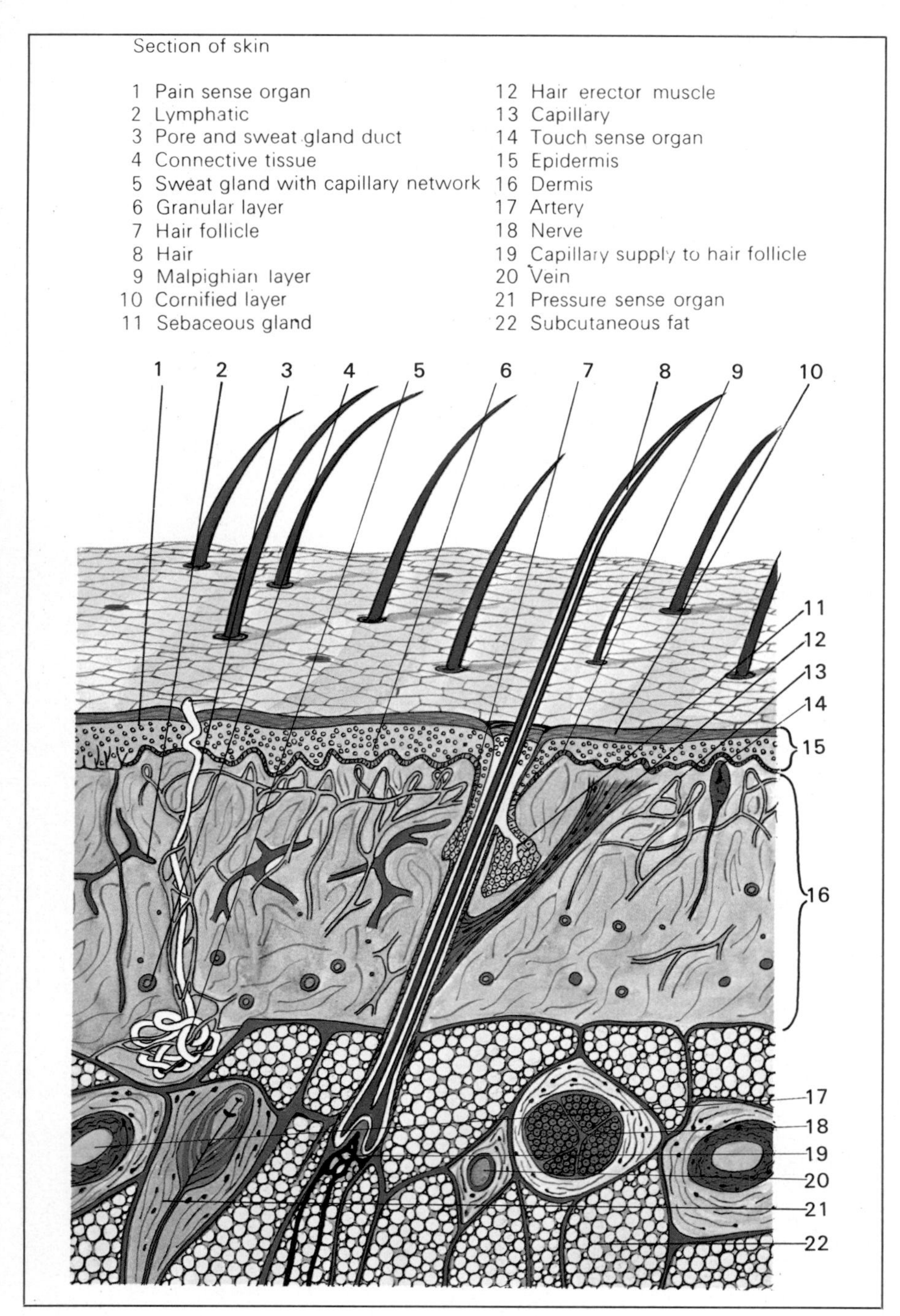

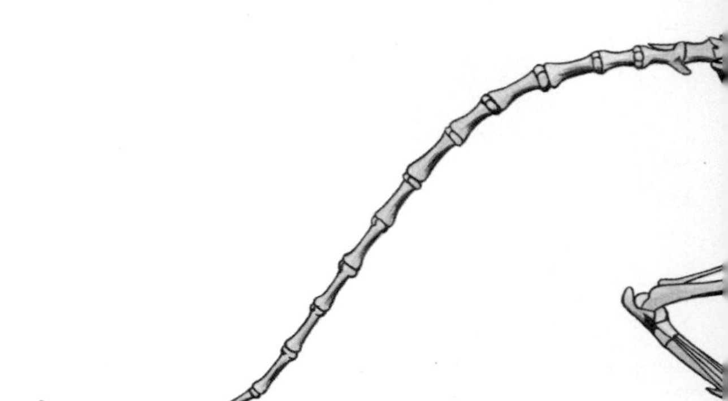

guard hairs form a protective covering. The spines of hedgehogs and porcupines, armour plates of armadillos and pangolins, the baleen plates in the mouths of whales and the horns of rhinoceroses are made from modified hairs.

The improved breathing and blood systems of mammals help to transport the extra fuel and oxygen needed to provide energy not only for keeping warm but also for activity. The secondary palate in the roof of the mouth provides an airway so that a mammal can continue to chew its food while breathing, and the diaphragm improves the regularity of breathing. A four-chambered heart, as in birds and crocodiles also, gives a better blood circulation.

The high body temperature increases the efficiency of the muscles and nerves. This, in turn, enables mammals to obtain the large amounts of food needed for them to maintain the high body temperature.

An obvious difference between the skeleton of reptiles and mammals is that reptiles sprawl with their legs out sideways, while the limbs of mammals are underneath the body to prop them clear of the ground and give more efficient locomotion. Variations in the bone structure of the limbs, especially in the hands and feet, have enabled mammals to specialise in fast-running, climbing, digging, flying and swimming.

A less obvious but very important difference lies in the jaws. A reptile's jaw is made up of several bones but only one remains in a mammal's jaw, the others having either been lost or become part of the ear mechanism. As mammary glands and hair are not preserved, it is the possession of a single jaw bone that is used by palaeontologists to identify a fossil as a mammal.

The teeth of mammals are set in sockets and most species have two sets: the milk-teeth, and their replacements, the permanent set. Most mammals, except for dolphins and some others, have several kinds of teeth.

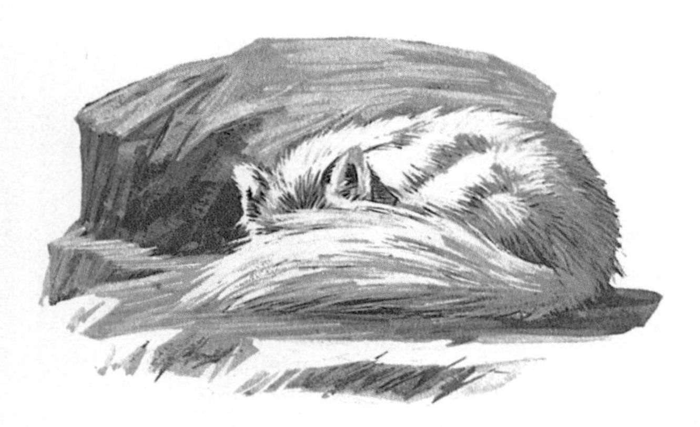

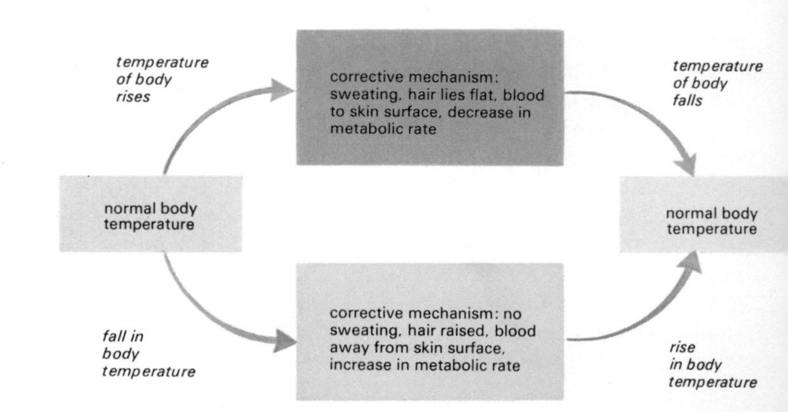

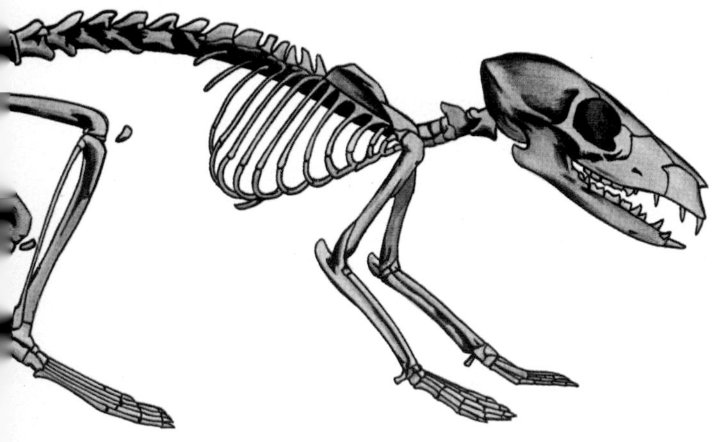

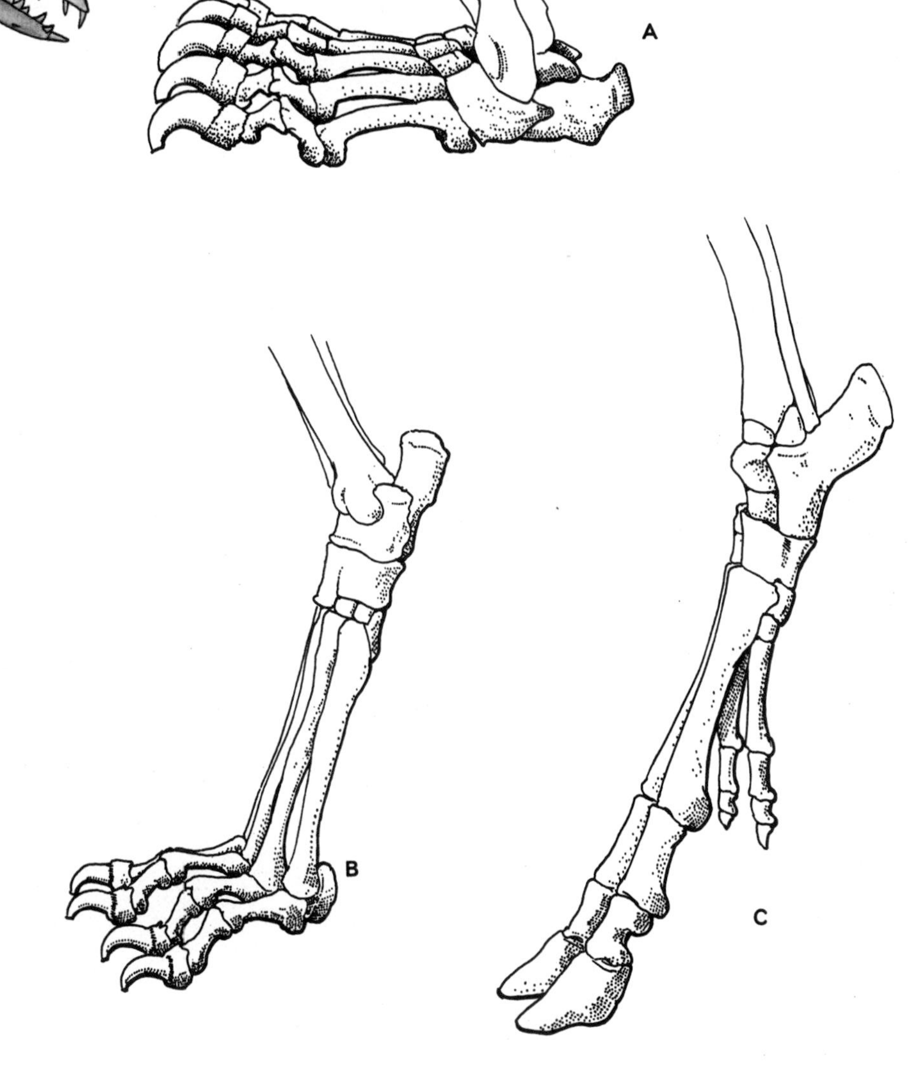

The incisors, or cutting teeth, are in the front of the mouth. Behind these come the canines, which are stabbing fangs in carnivores, followed by the cheek-teeth, the premolars and molars, which are for chewing and grinding. The number and shape of the teeth are used in the classification of mammals. This is very convenient because teeth are often preserved as fossils when the bones have disappeared. Teeth also reflect the diet and life style of mammals, and show whether they are carnivores, plant-eaters, insect-eaters and so on. Teeth often reveal a mammal's age either by the amount of wear or, more accurately, by growth rings, which are the equivalent of the rings in a tree trunk.

Mammals have large brains, with special development of the cerebral hemispheres and the cerebellum. The cerebellum is concerned with muscular co-ordination and balance, and the cerebral hemispheres are associated with greater powers of learning and adaptability of behaviour. Individual mammals are, therefore, better able to cope with novel situations and to adapt to changes in their environment than other kinds of animals.

Opposite top: **the skin of a mammal is a complicated structure. The hair helps to keep the body warm by insulation, while sweat glands help to cool it.**

Opposite bottom: **heating and cooling mechanisms work to keep a balance: when running, the wolf pants to keep cool; when asleep, the Arctic fox curls its bushy tail over its face to keep warm.**

Above: **skeleton of a tree shrew, a primitive mammal.**

Above right: **the main changes in other mammals are in the limbs: (A) plantigrade– for walking, as seen in bears; (B) digitigrade– as possessed by the agile carnivores; and (C) unguligrade– suited to running, exemplified here by antelopes.**

Right: **in the pangolin the hair has been transformed into flat armour plates.**

Mammals at night

It is fairly certain that the first mammals had a nocturnal way of life. As they were warm-blooded and maintained a high body temperature they could be active when the air temperature dropped after sunset. In this way they could avoid many of the reptiles, which would prey on them or compete with them for food.

The early mammals were insect-eaters, and they searched for food and found their way by smell, touch and hearing. The development of senses for life in the dark probably stimulated the development of the mammals' brains. Compared with the birds, which are mainly diurnal, most mammals do not have good eyesight. Those that use their eyes at night have a tapetum, a reflective layer behind the retina, to increase the eye's sensitivity. Light reflected by the tapetum causes the eye to shine at night.

Many modern mammals are still active by night. The monkeys, apes and man are exceptions. Some, like the cats and hoofed animals, can be active by day or night, but others, like bats, flying squirrels, the aardvark, hedgehogs, armadillos, bush-babies, gerbils and jerboas, are active almost exclusively at night. Yet even such nocturnal animals as bats are sometimes seen by day.

There are three main reasons for choosing to be active at night. The cool, moister atmosphere of the night may be more favourable, although this is not so important for warm-blooded animals except for those living in hot deserts. Darkness can give safety from predators, which could be important for small mammals – although there are many predators that hunt at night. For instance, persecution by humans has often been the cause for animals becoming more nocturnal. When lions were shot in the Kruger National Park in a misguided attempt to preserve other mammals, the survivors responded by hiding up during the day. The third reason is to make use of a particular food supply, for instance preying on other nocturnal animals, or to reduce competition for food by sharing out the 24 hours with other animals having the same feeding habits. For any species of mammal it is probable that a combination of all three factors could be operating, but this is a subject that has not received much study.

Bats

Bats are the only true flying mammals. The flying foxes, flying lemur and the marsupial flying phalangers and honey gliders have not mastered powered flight and can only glide. The bats are as efficient fliers as the birds and probably took to a nocturnal life to avoid competition with and predation by birds. The 950 species of bats include those that prey on fish, rodents, scorpions and even other bats, and the vampire sucks the blood of large mammals. Many species have an extremely efficient system of echolocation for finding insect prey. It includes the use of the Doppler shift, as in radar, which helps to distinguish a moving target from background objects. However, the fruit bats rely on large, sensitive eyes and even those with echolocation probably use their eyes for navigation.

Above: **bats, like this serotine, are very specialised nocturnal mammals. They have large throat muscles that aid the production of ultrasonic sounds, the echoes of which are heard by their well-adapted ears.**

Below: **bats are probably restricted to operating at night to avoid clashing with birds. Their characteristic feature is the wings, which are membranes attached to the limbs and tail. They are seen here in a pipistrelle.**

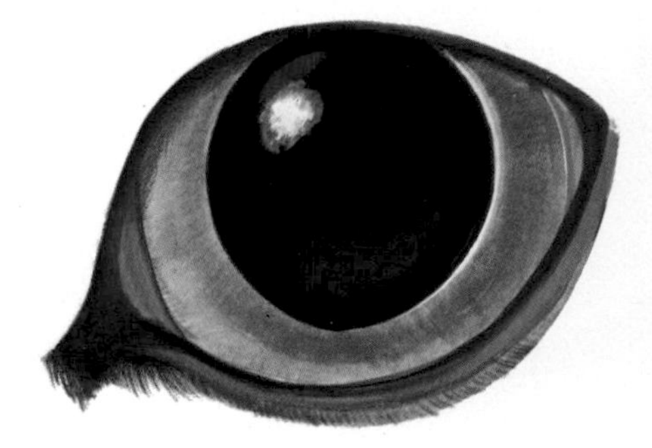

In dim light the pupil is exposed as much as possible to allow the maximum of light to pass to the retina.

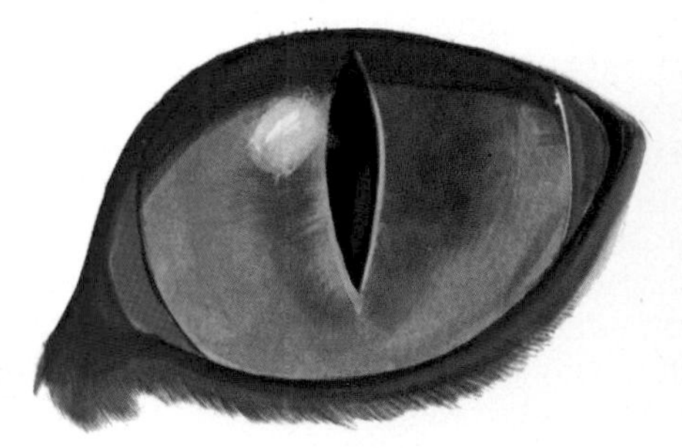

In bright light the pupil narrows to a slit.

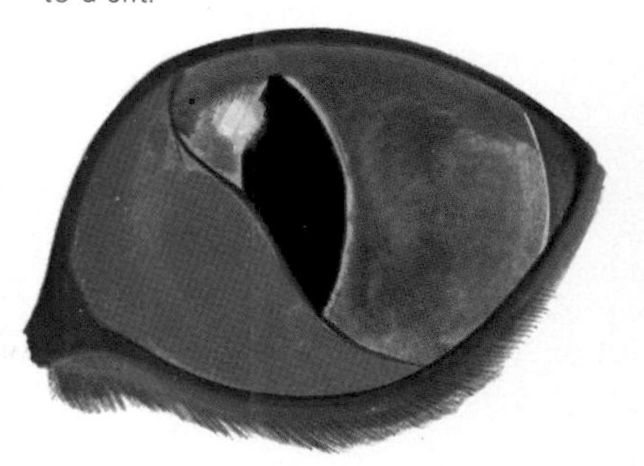

When a cat is unwell the third eyelid, the nictitating membrane, can be seen.

Opposite centre left: **although well-protected by their spines, hedgehogs come out under the cover of darkness to search for insects.**

Opposite bottom left: **the back and sides of the armadillo's body are covered by a skin thickened to produce a scaly shield as defence against predators. The three-banded armadillo, like the hedgehog, can roll into a tight ball when threatened, thus enclosing its unarmoured parts.**

Opposite bottom right: **the long-eared bats' outsize ears are used for detecting insects on leaves.**

The tarsier *(far left)* and the potto *(left)* are both nocturnal mammals which are not unlike the wider-eyed bush-babies but are unrelated. They inhabit the Philippines and Indonesia and insects form the main part of their diet.

Below: the bat skeleton shows how its arms are modified to become wings. The skull shows the sharp teeth, like a shrew's, for crunching insects.

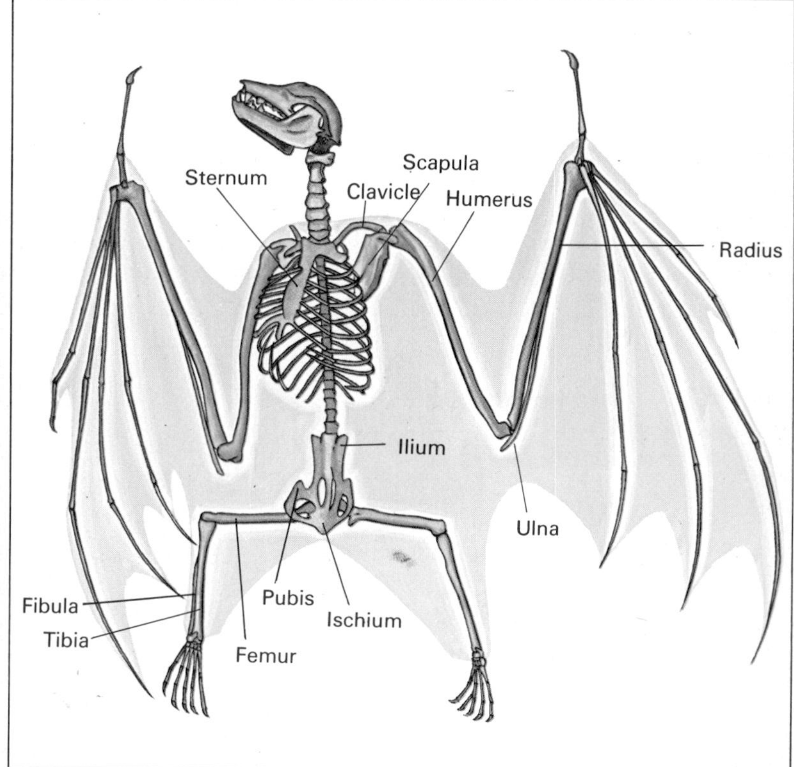

Reproduction

Bearing the young alive instead of laying eggs is a feature of the mammals, but it is neither a unique nor a universal character of them. Some reptiles, fishes and other animals practise forms of live birth, while one group of mammals, the monotremes, lays eggs. Nevertheless, the mammals characteristically prolong the care of their young both before and after birth so that their chances of survival are increased. In social species, such as lions and elephants, the offspring may remain with their parents for several years and raise their own young as part of an extended family.

The two monotremes, the platypus and the echidna, have several reptilian characters, including the way the embryos develop in the eggs. The platypus lays its two eggs in a burrow and they hatch in 10 to 12 days. The echidna lays a single egg and incubates it in a pouch on the belly. Their mammary glands do not have teats and the young monotreme licks the milk as it oozes out of the glands into the fur.

The marsupials or 'pouched mammals' give birth to their young at a very early stage of development. A kangaroo is born after only 40 days' development, when it is only two centimetres long. Its hindlimbs have hardly started to develop but the forelimbs are already fully formed, and it drags itself through its mother's fur and into her pouch. There, it takes a teat into its mouth and remains attached while its development is completed. The young kangaroo leaves the pouch for the first time when eight months old, but it returns to feed for another six months. Meanwhile, another baby has been born and is attached to one of the teats.

The majority of mammals are placentals, which means that the developing embryo is retained in the mother's womb, or uterus, and is nourished by food and oxygen carried in her blood. The embryo is linked by the umbilical cord to the placenta in which materials in the mother's blood stream are transferred to its own blood.

Some mammals, such as mice and human babies, are born in a helpless state. They are naked and have to be kept warm by their mothers, and new-born mice are blind and

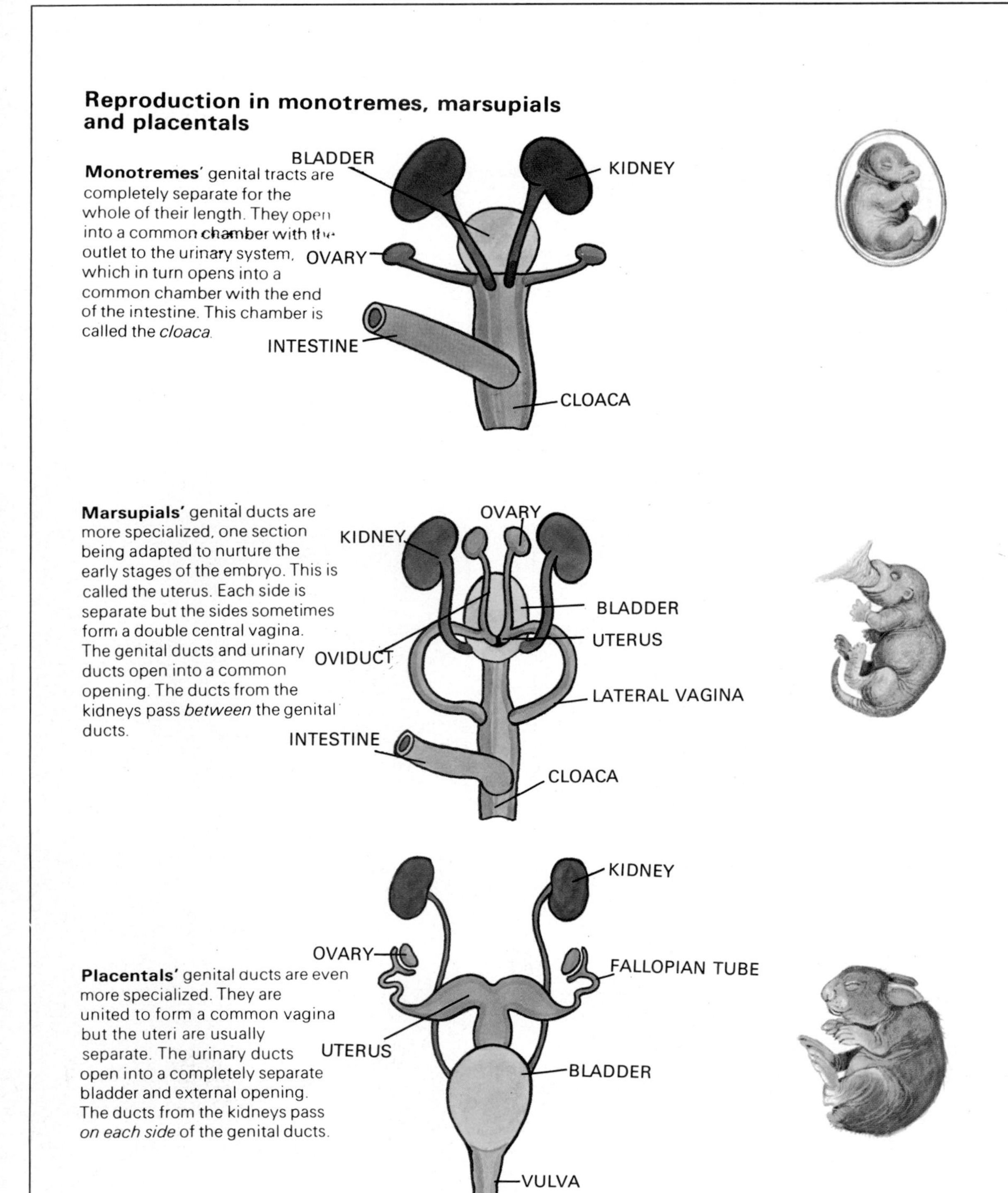

Reproduction in monotremes, marsupials and placentals

Monotremes' genital tracts are completely separate for the whole of their length. They open into a common chamber with the outlet to the urinary system, which in turn opens into a common chamber with the end of the intestine. This chamber is called the *cloaca*.

Monotreme young are hatched from an egg laid by the mother and incubated in a pouch. The pouch is temporary and only develops when needed. The young hatch in an immature state after a short period of development and spend a long time in the pouch.

Monotremes' mammary glands are located in diffuse patches under the ventral body wall. They open to the outside through multiple enlarged pores. There are no nipples.

Marsupials' genital ducts are more specialized, one section being adapted to nurture the early stages of the embryo. This is called the uterus. Each side is separate but the sides sometimes form a double central vagina. The genital ducts and urinary ducts open into a common opening. The ducts from the kidneys pass *between* the genital ducts.

Marsupial embryos develop in the early stages in the uterus. The young are born through the vagina in a very immature state with no hair, no eyes, limb buds only, and no true skin. They crawl into the pouch and attach to a nipple. The development is completed in the pouch. Pregnancy is very short.

Marsupials have discrete and definite mammary glands. The nipples are well developed and when in use are very elongated.

Placentals' genital ducts are even more specialized. They are united to form a common vagina but the uteri are usually separate. The urinary ducts open into a completely separate bladder and external opening. The ducts from the kidneys pass *on each side* of the genital ducts.

Placental young develop in the uterus, nourished through the placenta. They are born in a well developed state, usually with hair. They are suckled at regular intervals by the mother but do not attach to the nipple. There is no pouch. Pregnancy is longer.

Placentals have definite mammary glands with nipples not so well developed.

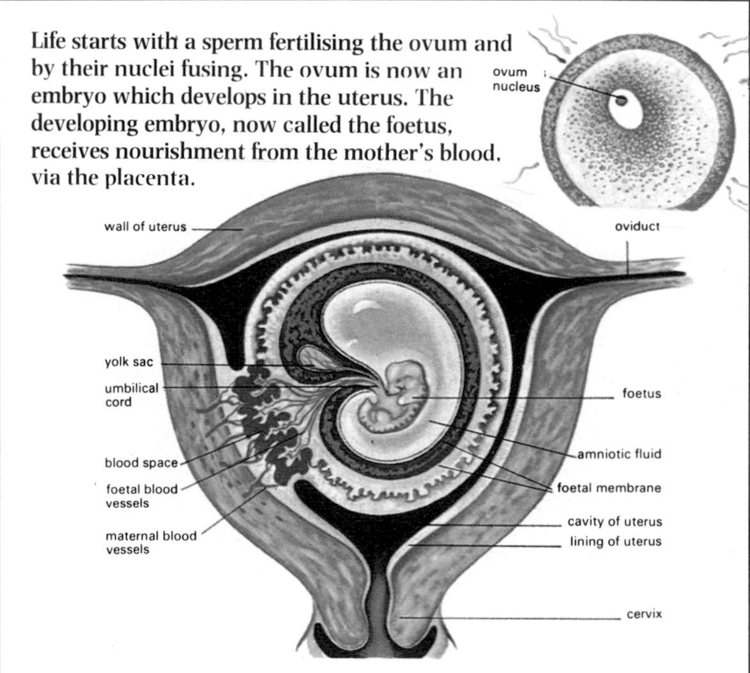

Life starts with a sperm fertilising the ovum and by their nuclei fusing. The ovum is now an embryo which develops in the uterus. The developing embryo, now called the foetus, receives nourishment from the mother's blood, via the placenta.

deaf. Other mammals, such as horses, cattle and giraffes, are advanced enough to stand within a few minutes of birth, and dolphins have to swim immediately to the surface to breathe.

The number of young born at one time in a litter ranges from (usually) single human babies and elephants to over 20 for house mice and brown rats, although average litters are smaller. As a general rule the litter size is related to the length of the animal's life. Mammals which are likely to live for only one year or so produce large litters and may breed several times in a year. House mice can breed continuously producing up to 10 litters a year. Long-lived species bear one or two young at a time and elephants, which live to 70 years, bear a calf only once every four or more years.

Above right: **the pouch of the female Virginia opossum, one of the American marsupials, showing the young at an early stage of development.**

Above: **the platypus, one of the egg-laying monotreme mammals. The young are licking milk as it oozes from the milk glands.**

Below: **wildebeest with new-born suckling calf. The calves can walk within a few minutes of birth, and run after a few hours, which lessens their chances of being taken by predators.**

Evolution

An ancestor of the horse and deer, ***Phenacodus*** still had the long body and tail of carnivores. Its small hooves and broad cheek-teeth identify it, however, as an ungulate.

Uintatheres (in the background) and ***Eohippus*** were among the early herbivorous mammals that evolved rapidly and roamed the land 66 to 26 million years ago.

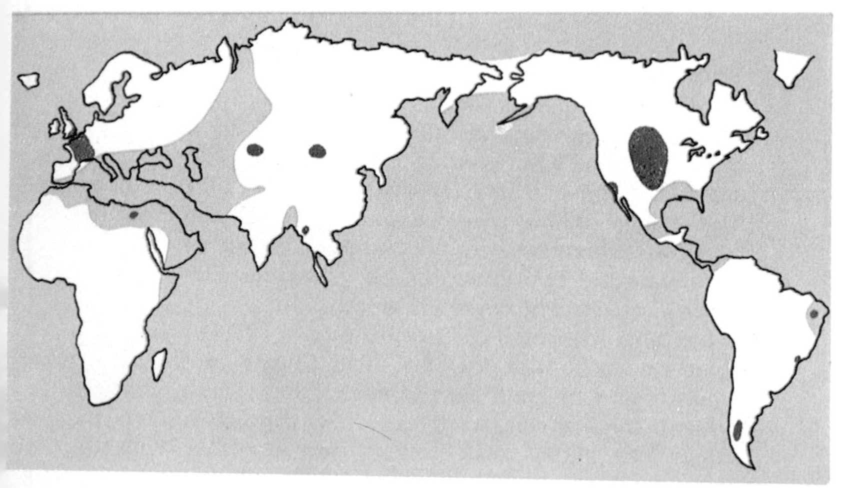

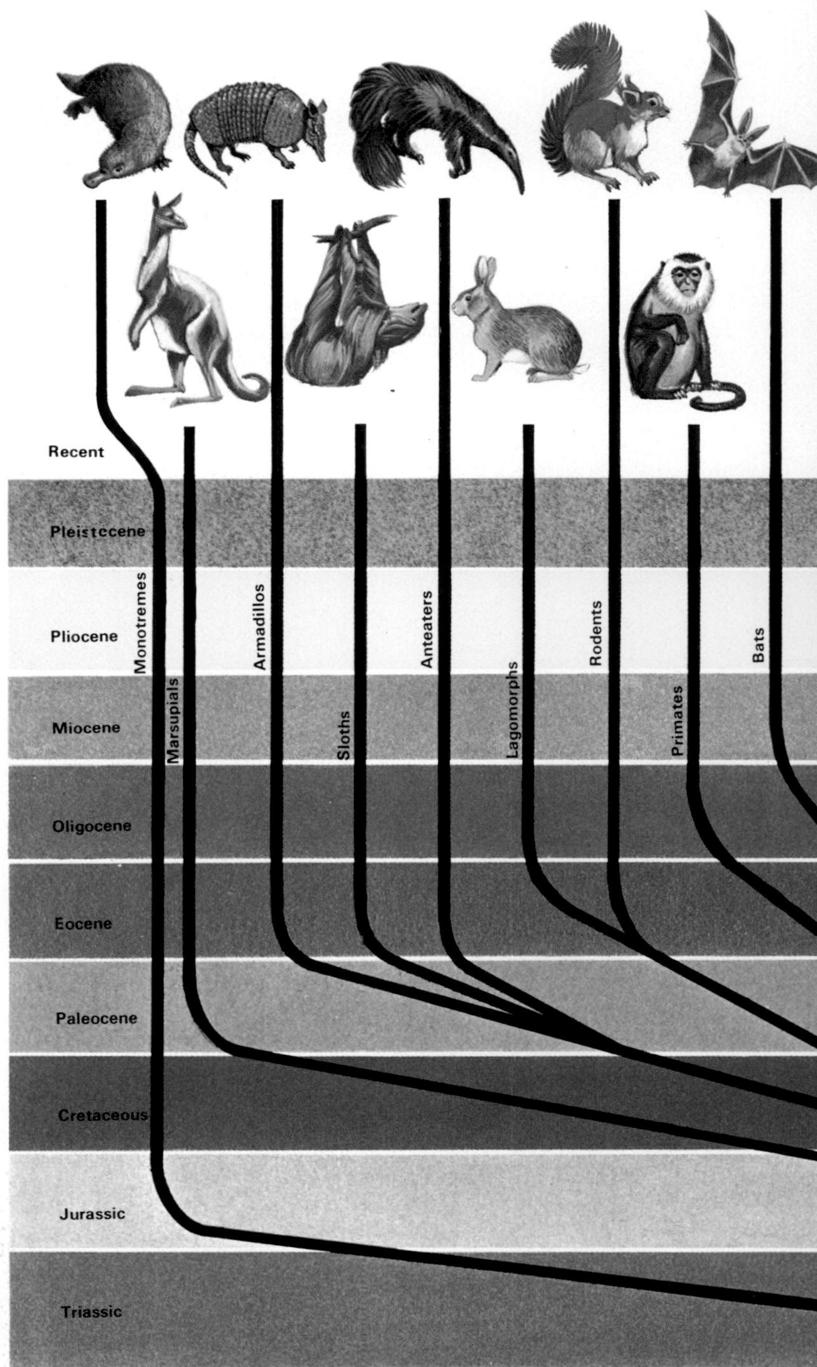

Above centre: map of land masses during the Lower Tertiary period when mammals first became abundant. Red indicates main regions for fossil finds.

Above: the shrew-like ***Morganucodon***, the first known mammal, lived at the same time as the early dinosaurs.

Right: a detailed scheme for the evolution of mammals. It shows the age of each group and its relationship with other mammals.

The first known mammal was an animal called *Morganucodon*. It was only a few centimetres long and probably looked and behaved rather like a shrew. It lived about 200 million years ago in the Triassic Age, at a time when the age of the dinosaurs and other giant reptiles was only just beginning.

Because the soft parts of the body are rarely fossilised, the evolution of animals is discovered mostly from the remains of their skeletons. The ancestry of the mammals can be traced through a series of reptile fossils which increasingly became more mammal-like. This 'paramammal' line started among the earliest reptiles, first as mammal-like reptiles and then as reptile-like mammals. They developed warm-bloodedness and became more active; a secondary palate in the roof of the mouth improved their breathing; and their jaws and ears became more efficient.

From a position of dominance in the Permian Age, the paramammals rapidly became extinct when the dinosaurs evolved from aquatic ancestors and spread over the land during the Triassic. *Morganucodon* and the other true mammals survived, however, probably because they were active at night when the large dinosaurs were asleep.

About 64 million years ago, at the end of the Cretaceous, the ruling reptiles suddenly disappeared. The reason is not fully understood, but the mammals must have hastened their end through competition for food. One dinosaur needed as much food as thousands of shrew-like mammals, so when dinosaurs were finding food scarce the mammals could still flourish. Their warm-blooded bodies and quick-thinking brains enabled them to cope easily with changing conditions.

With the giant reptiles removed from the scene, the mammals could come out during the day, but they had to compete with the birds, which had also been evolving during the Age of Reptiles. At this time there were already primates, opossums, shrews, hedgehogs and the rat-like multituberculates and flesh-eating miacids which later became extinct. They were soon joined by the first bats and rodents, the earliest carnivores and the plant-eating, pig-sized condylarths that probably gave rise to the hoofed mammals. The largest of these, and the largest land mammal ever to have existed, was *Baluchitherium*, which was five metres tall at the shoulders and weighed 16 tonnes.

Many of the early mammals looked very strange, especially those with horns or antlers. *Uintatherium* was like a rhinoceros with small tusks and several pairs of knobby horns. *Synthetoceras* had a pair of horns between the ears and another forked horn on the snout.

Among these odd mammals, the fossil record shows two very clear lines. The horse developed from the fox-sized *Hyracotherium*, which ran on its toes and eventually the horse came to stand on the nail of one toe – the hoof. The first elephant, *Moeritherium*, had neither trunk nor tusks, but its many descendants tried many shapes of tusks.

The first primate, *Purgatorius*, was alive during the Age of Reptiles, but modern monkeys and apes came much later. The last to appear was man, *Homo sapiens*. The first man-like ancestor, *Ramapithecus*, lived 20 million years ago and newly-found fossils have revealed a bewildering range of man-like apes and ape-like men that succeeded it until modern man eventually emerged 20 000 years ago.

Below: **four of the main kinds of elephant, ending with the woolly mammoth which was related to modern elephants.**

Modern mammals

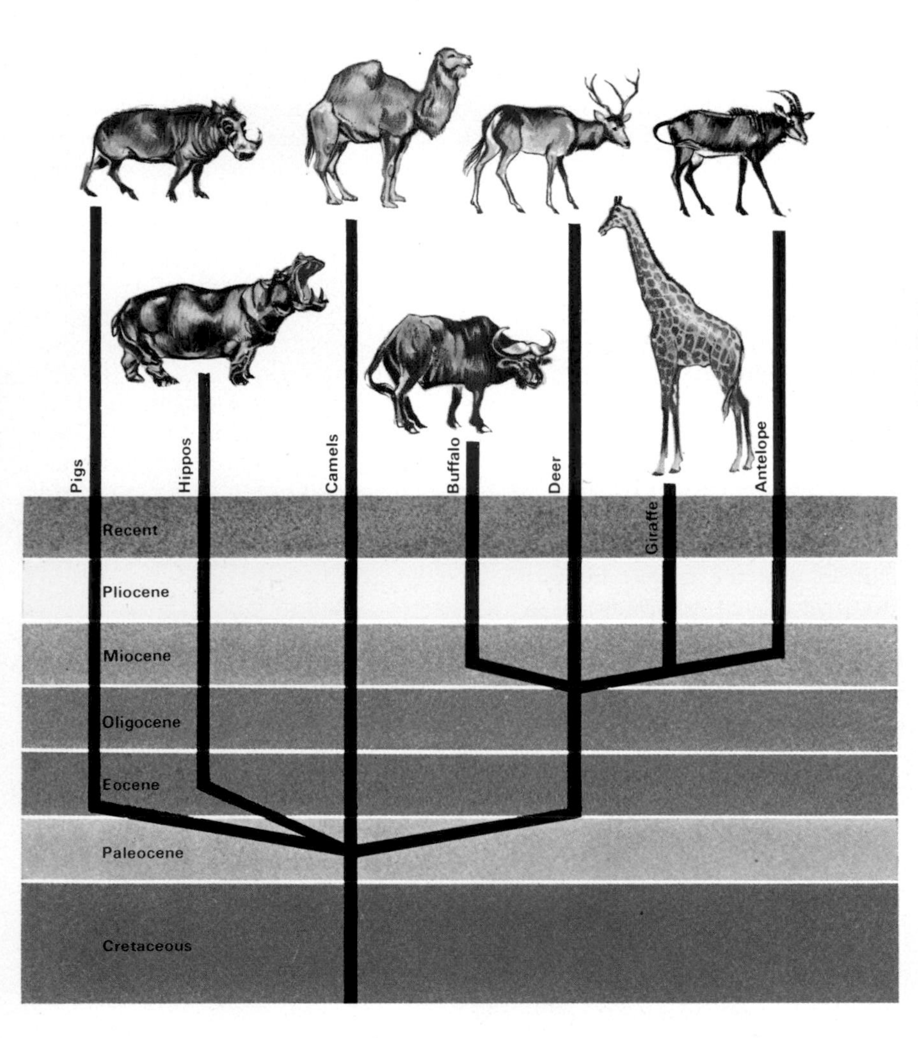

Every animal and plant has a scientific name based on Latin words, which is recognised by people of all nationalities. This avoids misunderstandings for common names can be very confusing. For instance, the deer that the British call the red deer is known as the wapiti or sometimes the elk in North America. Yet elk is the European name for a much larger deer which the Americans call the moose. Scientists call the first deer *Cervus elaphus* and the larger deer *Alces alces*.

Each scientific name is composed of two parts – the genus name and the species name. A genus may be composed of several species, all of which are closely related. *Cervus elaphus*, for example, has the close relatives *Cervus unicolor* and *Cervus eldi*. Several genera are grouped into a family – all deer are in the family Cervidae – and families are grouped into orders – the deer are in the order Artiodactyla. The various orders are united in the class Mammalia. This classification is called a natural system because it shows the relationship between animals and the probable course of their evolution. The chart shows how the even-toed hoofed animals are related to each other.

There are 20 orders of living mammals. All are placentals except the monotremes and marsupials. Some orders, such as the Carnivora and Primates, are familiar; others like the flying lemurs and the tree shrews are rarely heard of. Sometimes the links between different orders are known, as with the seals, which are carnivores that took to the sea, but it is not at all clear, for instance, which are the closest relations of the cetaceans or the aardvark.

The mammals have adopted an enormous variety of ways of life. They can fly, swim, run, burrow and climb, and they feed on every kind of animal and plant. The result is that mammals have invaded every part of the globe, from deserts to rain forests. Moreover, their intelligence enables them to adapt easily and quickly to new conditions.

CLASS MAMMALIA

Order	
Monotremata	platypus, echidnas
Marsupialia	pouched mammals: kangaroos, wallabies, koala, possums, bandicoots
Insectivora	moles, shrews, hedgehogs
Tupaioidea	tree shrews
Dermoptera	flying lemurs
Chiroptera	bats
Primates	lemurs, lorises, tarsiers, monkeys, apes, man
Edentata	sloths, anteaters, armadillos
Pholidota	pangolins
Lagomorpha	rabbits, hares, pikas
Rodentia	rodents: rats, mice, squirrels, porcupines, beavers, gerbils
Cetacea	whales, dolphins, porpoises
Carnivora	cats, dogs, bears, hyaenas, weasels, otters, pandas, civets, racoons
Pinnipedia	seals, sea lions, walrus
Tubilidentata	aardvark
Proboscoidea	elephants
Hyracoidea	hyraxes
Sirenia	dugong, manatees

Ungulates or hoofed mammals	
Perissodactyla	odd-toed ungulates: horses, zebras, tapirs, rhinoceroses
Artiodactyla	even-toed, or cloven-hoofed, ungulates: pigs, camels, hippopotamuses, cattle, sheep, goats, deer, antelopes

Opposite top left: **a simplified view of the evolution of the even-toed mammals, the Artiodactyla. The pigs and peccaries, the hippopotamuses, and the camels and llamas, split from the rest at an early stage. Giraffes and deer are more closely related to the buffalo and antelope group.**

Opposite top right: **giraffes and okapis have horns which are bony outgrowths covered by skin, unlike either deer or antelopes.**

Opposite bottom: **the fallow deer and its relatives shed their bony antlers once a year, so they should not be confused with the similarly built antelopes which carry permanent horns.**

Top: **the aardvark is a mammal on its own, but it has anteating habits like the anteaters, pangolins and echidnas.**

Right: **manatee and calf. Together with the dugong, these animals are sea-cows or sirenians. They have evolved the same hairless streamlined bodies and paddle-shaped tails as the whales, although they are shallow-water vegetarians.**

Australia

There is an overwhelming difference between the mammals of Australia and Asia. Asia is the home of carnivores, hoofed mammals, primates and other placental mammals. Australia is the home of kangaroos, possums, koalas and other marsupial mammals, and there are only a few native rats and bats to represent the placentals. The two continental landmasses are linked by the mass of islands which includes the Philippines, Indonesia and New Guinea and the boundary between the two groups of mammals, as well as of birds and reptiles, runs through them, with New Guinea, the Philippines and Celebes on the Australian side. The boundary is known as Wallace's Line, after the 19th century English naturalist who first noticed the sudden change in animal species.

The reason for the difference between Asian and Australian animals is that the marsupial mammals reached Australia before it became separated from the rest of the world and the placental mammals arrived too late to get across to it. But why one group of mammals got there and the other did not is something of a mystery.

Australia is famous as the home of the marsupials, and it is not usually realised that it has almost as many placental species. However, they are of only two types which managed to leap-frog across the islands from Asia. Bats were able to fly across the sea between the islands and rodents drifted across, perhaps on logs. There are 119 species of marsupials, rather more than half the total of Australia's mammals, and, although about the same number of rodents reached New Guinea, only half as many reached Australia.

Since European settlers arrived in Australia two centuries ago there have been many changes in its mammals. Cats, foxes, rabbits, mice and rats have been introduced. They have either preyed on the native mammals or have competed with them for food. Cutting down the forests and turning the countryside into farmland has destroyed the homes of many species so that some, like the thylacine or marsupial 'wolf' and the Alice Springs mouse, are now extinct and others are very rare. Very few have been able to compete on equal terms with the newcomers.

The wildlife of New Zealand, separated by 1600km of ocean, has little in common with Australia and there are only two native mammals. Both are bats which must have been blown by the wind across the sea.

The two strangest Australian mammals are the platypus and the echidnas, or spiny

Above: **a party of kangaroos rests in the shade. The largest of marsupials, they are plant-eaters which became unpopular when they started to compete with sheep on pastures.**

Left: **echidnas, or spiny anteaters, are egg-laying monotremes like the platypus. The one on its back shows the 'incubation groove' where the egg is brooded.**

Above left: **the swamp wallaby is a widespread species. Wallabies are no more than small species of the kangaroo.**

Top right: **the Tasmanian devil is a flesh-eating marsupial, once found throughout Australia, but now confined to Tasmania.**

Centre right: **the koala lives only where it can find eucalyptus, or gum, leaves to eat.**

Left: **the hairy-nosed wombat is a badger-like marsupial which lives in burrows.**

Bottom right: **four tree-dwelling marsupials belonging to the phalanger family. They are assisted by prehensile tails.**

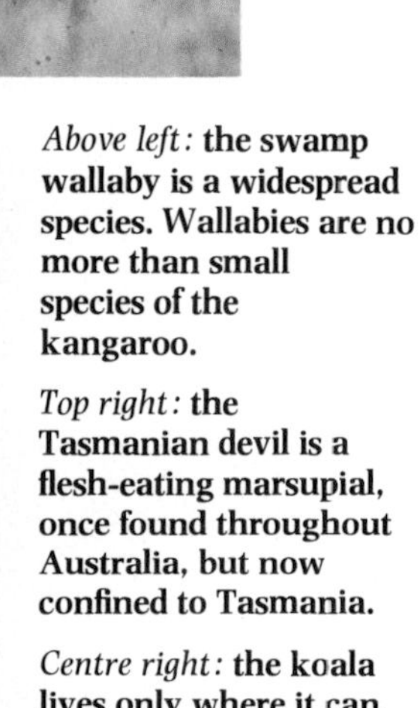

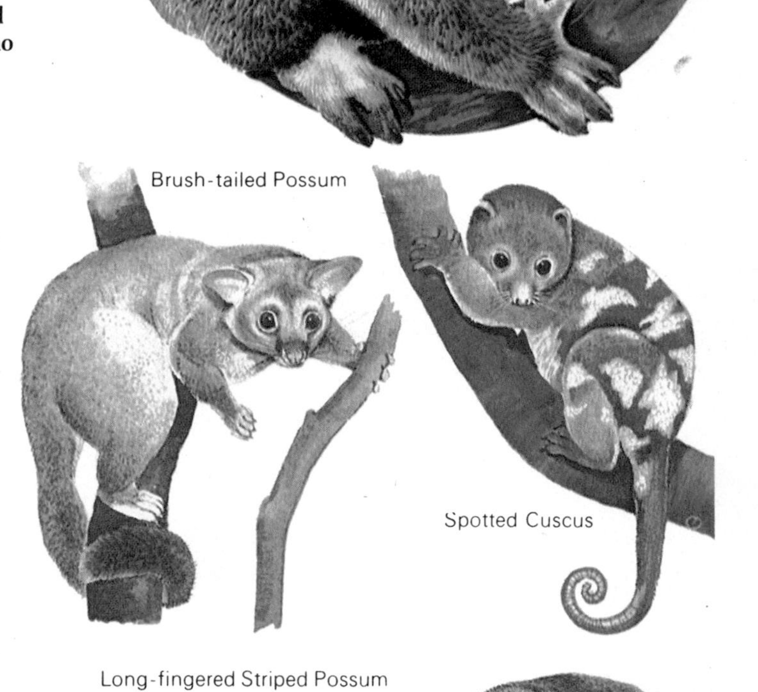

anteaters. These are monotreme mammals which lay eggs. The platypus lives in rivers where it feeds on insect larvae and other invertebrates. It closes its ears and eyes when underwater and relies mainly on the very delicate sense of touch in its leathery bill for finding food. The two echidnas look rather like long-nosed hedgehogs, and their homes range from wet tropical forests to dry deserts. The Australian species eats mainly ants and termites, which it gathers by breaking open their nests, inserting its snout and wiping up the insects with a long, sticky tongue. The New Guinea species eats mainly earthworms.

Marsupial variety

Some groups of animals show a diversity of form and habits among their members. Although related, they have become adapted to take up different ways of life. A mammal, for instance, may grow long legs for running, or long hindlegs for jumping, so that it can move about quickly to find food or escape enemies. Its relatives may have developed strong, short limbs for digging, paddle-shaped limbs for swimming, or grasping limbs for climbing. This process is called adaptive radiation and it occurs when a group of animals arrives in a new environment or a new place where there is no competition from other animals.

The marsupial mammals of Australia are an example of animals that have expanded in an unoccupied space. Without competition from the placentals, which did not reach Australia, the marsupials were free to evolve species specialised in many ways of life. Some of the Australian marsupials have remarkable similarities with the placentals in the rest of the world. This is called parallel evolution.

The most familiar Australian animals are kangaroos and wallabies. They are the marsupial equivalents of the ruminating or cud-chewing placentals, such as deer and antelopes. In common with these placentals, kangaroos have a stomach divided into compartments where protozoans digest cellulose into substances which the mammals can assimilate. The bounding gait of kangaroos and long, slender legs of ruminants are different means of achieving the same end of fast and efficient locomotion.

There are 50 species of kangaroos, from the 0·5kg musky rat kangaroo to the 90kg red kangaroo. In the tropical forests of New Guinea and northern Australia the tree kangaroos have lost the ability to bound, and instead they clamber among the branches.

The chart shows the variety of Australian marsupials and the parallels among the placentals. Note how the marsupial mole has the same strong, shovel-shaped limbs and streamlined body for fitting into a burrow as has the true mole, and how the hunting thylacine or 'Tasmanian wolf' is remarkably similar to the true wolf. The Australian marsupials have failed to evolve a true flier, perhaps because bats had arrived, or an aquatic form, perhaps because the platypus filled that gap.

In South America there has been a second adaptive radiation of marsupials. Not all placental types reached South America and some of its marsupials parallel placentals elsewhere. The South American marsupials, for instance, include the equivalents of the African lorises and pangolins. There is also an aquatic marsupial, the yapok or water opossum, which can close the mouth of its pouch when swimming so that the young do not drown.

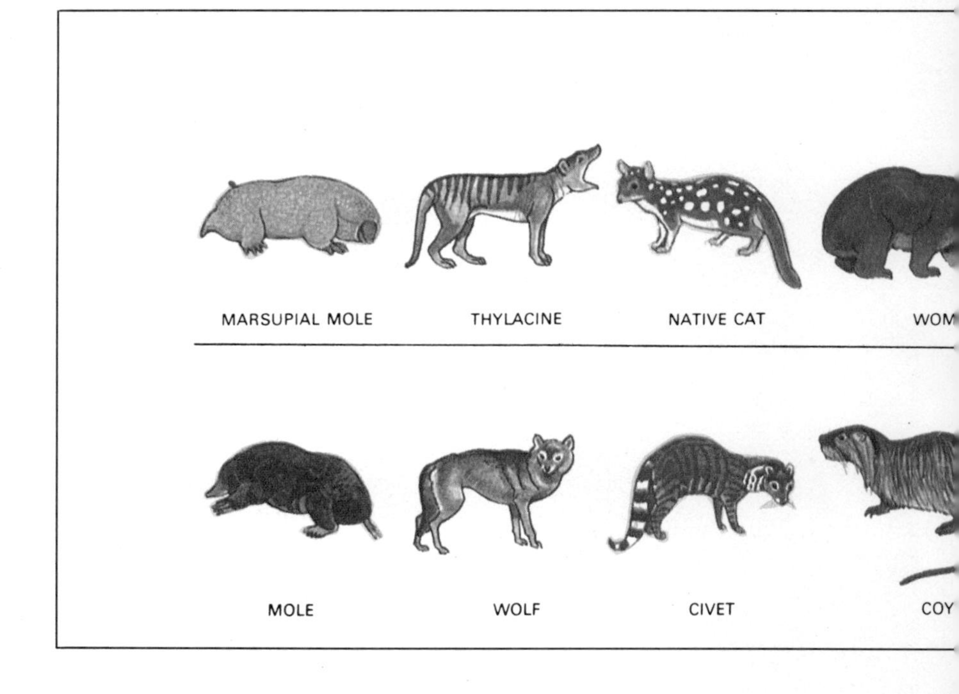

Right: **parallel evolution between marsupial and placental mammals. There are similarities in the way of life and appearance in species of the two groups; although a kangaroo does not look like a roe deer, both are grazing animals which fill similar habitats.**

Below: **the Virginia opossum is the only marsupial in North America, where it survives among the placental mammals.**

Below right: **the great glider possum, one of the 'flying phalangers' which can travel considerable distances through the air on flaps of skin stretched behind fore and hind limbs.**

A 'flying phalanger' in mid-flight. Its 'wings' give it a range of about 100 metres and its tail helps it to steer and balance.

Lumholtz's Tree Kangaroo

Tree kangaroos live in the forests of northern Australia and New Guinea. The hands and feet are adapted for grasping branches but the tail is not prehensile— it is used for balancing.

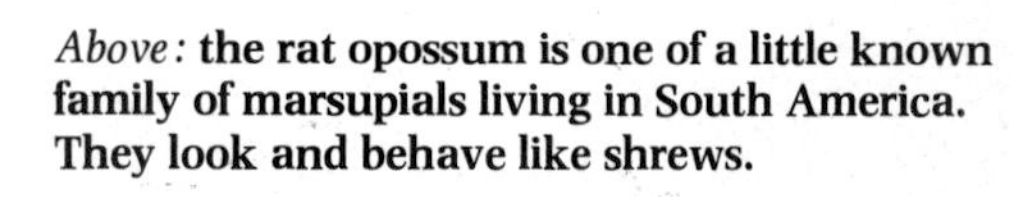

Above: the rat opossum is one of a little known family of marsupials living in South America. They look and behave like shrews.

Left: the numbat or banded anteater catches ants and termites with its long tongue. It has fifty small teeth for crunching them.

Tundra

The sun disappears in the Arctic winter, leaving the countryside bitterly cold. It returns in spring, and during the summer months it never sets. The snow melts, the surface of the ground warms up but parts of the tundra remain a complete desert, because there is so little moisture that nothing can grow. Only where water from the melting snow is trapped by the uneven terrain can a rich profusion of plants and animals flourish.

The few mammals living on the tundra are those that have adapted to survive the long, hard winters and take advantage of the short summers. Unlike the birds, they do not migrate to escape the winter, with the exception of the caribou, which head southwards in large herds along traditional trails to the shelter of the forests.

The musk ox, the other large tundra grazer, relies on a coat of very fine, dense underfur, 15cm thick, protected by a long, shaggy overcoat. The musk ox conserves its energy by standing quietly and huddling with the rest of the herd. Although relying mainly on its fat, it can get some sustenance by digging through the snow. When attacked by wolves the herd gathers into a tight bunch, called a *karre*, and faces the enemy. If a wolf comes too close, a musk ox will charge it, but retreat again to the *karre* before the wolf can cut it off from the group.

Lemmings live in underground burrows during the summer and excavate tunnels and nests under the snow in winter. This protects them from the worst of the weather and allows them to continue feeding. The collared lemming grows extra claws on its front feet to help in digging. Arctic hares also burrow into the snow in search of food and shelter.

The herbivorous mammals and, in summer, the birds provide a livelihood for carnivores. The stoat is an exclusive hunter of live prey but the Arctic fox will eat anything including carrion (and insects) and will follow polar bears onto the frozen sea to eat scraps from their kills. The winter is a lean time for polar bears and they spend most of their time sleeping, but not hibernating, in dens under the snow. They emerge in spring to feed on the new crop of seal pups.

Many Arctic mammals have white fur. Lemmings, foxes and stoats turn white in

Left: **the polar bear is the super predator of the Arctic. Its main food is seals, which it catches on the ice, but it also comes on land to eat other mammals and birds, or even to scavenge at refuse tips.**

Below left: **Arctic hares in their winter coats. The coats are a good camouflage against predators and keep them warm through the winter months.**

Below: **musk oxen are protected from the coldest weather by a thick coat of fine wool. They find a little to eat in winter by digging in the snow.**

In the thick white fur of winter, an Arctic fox curls up in the snow to conserve energy.

winter and polar bears and Arctic hares are white throughout the year. Among the birds, ptarmigans turn white in winter, while snowy owls and gyrfalcons are white all the year. White fur or feathers is a good camouflage but, in addition, it helps the animal keep warm. Energy from the sun penetrates white fur to warm the body and the fur then keeps the heat in.

Below: **tundra country is barren stony ground, with many pools, but there are patches where low plants grow well and provide food for animals, from mosquitoes to musk oxen.**

Taiga

To the south of the tundra there lies a belt of coniferous forest that stretches round the world. It is called taiga after a Russian name for the marshy forests of Siberia. Like in the tundra, the same or very similar animal species are found throughout the taiga, there having been free movement across the Bering Strait in past times. The winters are severe in the taiga but the trees give shelter and the warmer summers give a rich harvest of berries, seeds and nuts to support mammalian life.

Mice, voles, lemmings and hares inhabit the ground between the trees, while the squirrels and porcupines (in America) live in the trees. The flying squirrels have developed the leaping ability of the squirrel family to become expert gliders. They have a membrane of furry skin stretching between the forelegs and hindlegs, with extra support given by a spur of cartilage on the wrist. A flying squirrel can control the tension in the membrane and, aided by the furry tail, it can steer a course through the trees. Flying squirrels are also unusual in being nocturnal, presumably to avoid diurnal birds of prey.

The larger herbivores are represented by the moose or elk, the largest of all deer, which prefers wet swamplands, and reindeer.

The taiga is the land of the trapper because, preying on the plant-eating mammals, as well as on the numerous birds that take advantage of the seed crops, fruits and insects, there is a variety of fur-bearing carnivores. There is only one cat, the short-tailed, tassle-eared lynx, but there are bears, foxes and wolves and, above all, the mustelids, the members of the weasel family. Many of these animals are now rare because of the efforts of trappers, or of farmers protecting their livestock.

The mustelids range from the weasel, which is small enough to hunt rodents down their tunnels, to the powerfully built wolverine. Also called the glutton, the wolverine is a hyaena-like animal with strong jaws. It travels vast distances in search of prey, and in winter its broad feet enable it to run over snow and catch the deer, which founder on their slender hooves.

At one time the brown bears were divided into six species and many subspecies, but it is now realised that there is only one species, with regional varieties, living round the world. The grizzly, for instance, is only a large variety of the brown bear. Now that

Left: **the largest living deer, known as the moose in North America and the elk in Europe and Asia. It likes bogs and pools in the taiga forests.**

Below left: **the grizzly bear, a race of the brown bear, is an adaptable animal feeding on anything from berries and wild honey to deer and fish, but it is extinct in many parts of its old range.**

most of the northern deciduous forests have gone, the brown bear is found mainly in coniferous forests, where the autumn berry crop enables it to put on a thick layer of fat for the winter.

Populations of animals rise and fall naturally, but in the Arctic there are often violent changes. In a good year mice, lemmings and voles breed rapidly. Then the food runs out and large numbers of the animals die of starvation. Populations of small rodents reach a peak every four years, and snowshoe hares peak every 10 years. The fortunes of the predators follow those of the prey animals. The predators breed well when there is plenty of food but they starve or move away after a population crash. Hunters' fur returns in Canada show how numbers of lynxes follow the population cycle of snowshoe hares, their favourite prey.

Right: **the lynx once lived throughout much of North America and Asia but it has been persecuted and now survives only in a few places, including parts of the taiga.**

Centre: **a grizzly bear defending a deer carcase from marauding wolves. Deer are vulnerable in deep snow but predators often go hungry.**

Below: **the taiga in winter. Conifers survive accumulations of snow without their branches breaking and they form important refuges for deer and other animals.**

Bottom right: **reindeer have been really domesticated in Europe's northern latitudes and are herded throughout the Arctic Circle like cattle. Domestic reindeer have been recently introduced into North America.**

Deciduous woodlands

Before the time of the Industrial Revolution there were huge tracts of forest covering most of Europe, eastern Asia and eastern North America. These forests grow in moist temperate climates and, because their underlying soil is fertile, they have always been favoured places for human settlement. As populations grew and needed land for crops and timber for building and fuel, so the forests dwindled and now only scattered remnants survive. Many of the larger mammals also have gone: wild cattle and horses, bears, wolves, boar and deer have either lost their homes or been hunted by the increasing human population.

A deciduous wood consists of three layers: the canopy, the shrub layer and the herb layer. The canopy is formed by the foliage of the tall trees – oak, hornbeam, ash, maple, poplar and lime. The shrubs are smaller trees, such as hazel, cherry, holly and privet, and they form an understorey. The canopy and shrubs are the home of the squirrels and their predator the pine marten, the dormice in Europe and the opossum and porcupine in America. Bats hawk between the trees, and the long-eared bat hovers to pick insects and spiders off leaves.

These mammals live in trees and nest or roost in holes or construct their own nests of sticks and grasses. They are joined by visitors from the ground, like the wood mouse of Europe, the deer mice of America, stoat,

Part of a forest which once covered much of Europe. Where woods have survived it is usually because they have a purpose. Beechwoods, like the one shown here, were needed for the furniture industry.

The red fox is one of the most successful predators. It will eat almost anything and survives even when the woodlands have been cut down.

Above: **the pigmy shrew, only 6cm long, lives on the woodland floor, where it searches for insects and other invertebrates.**

Right: **wood mice are also called long-tailed field mice and they can be found in open country as well as in woods. They live in burrows and gather stores of nuts and berries in the autumn.**

A mole pokes its head out of its burrow. Moles are most obvious from their 'hills' in fields but they would have been largely woodland animals before the trees were felled. They are common wherever the soil supports plenty of earthworms, their main food.

weasel and wild cat. Wood mice make hoards of nuts and berries in holes or old birds' nests. Squirrels do not gather hoards but bury their nuts singly. They do not remember where they buried them but rediscover them by smell.

The ground is the home of voles, mice and chipmunks. The last are ground-living relatives of the squirrels. These mammals feed on low-growing plants or fallen fruits and seeds. They often have extensive runways through the matted herbage and nests under logs and boulders, which give them some protection from their many predators. The mice dig underground burrow systems, which can be mistaken for the underground workings of moles. These animals eat some invertebrate animals as well as plant food, but living prey is the main diet of the insectivores – the shrews, moles and hedgehogs.

Shrews are mouse-like with pointed snouts. They eat any small animals, vertebrate or invertebrate, which they can overpower, and some species eat quantities of seeds. They are extremely active and have alternating periods of foraging and resting every hour or so throughout the 24 hour day. The much larger hedgehogs of Europe and Asia are strictly nocturnal and are protected from their enemies by their coat of spines. When first alerted, a hedgehog 'freezes' and erects its spines. Further disturbance makes it curl up and pull the spiny coat around it to make a ball. The moles are insectivores which live in underground tunnels ranging from over one metre deep to so close to the surface that a visible ridge is made. The tunnels act as a trap into which worms and insect grubs fall and are picked up by the patrolling mole.

Three large omnivores are supported by the forests of the Old World. The brown bear survives only in the remotest places and is now largely confined to mountains and the taiga forests. Wild boar also have gone from much of the old range. Both are formidable animals that were hunted for sport and persecuted because of the danger they posed. Wild boar are the ancestors of domestic pigs and show the same ability to eat almost anything. The flat disc of the snout is a sensitive organ of touch and also a strong plough for rooting through the soil in search of earthworms, grubs, bulbs and tubers. The badger, a member of the weasel family, has survived where the other two have disappeared. It can live in small woods, even in open country and on the outskirts of towns.

A natural forest has glades and clearings where the increased low growth makes a good feeding ground for herbivores. Such places attract woodland deer. The red deer is a woodland animal over most of its range,

Above: **the red deer, which grows a magnificent set of antlers each year, lives throughout the woodland range.**

A doe and fawn white-tailed deer, the most widespread deer of North America. Although principally browsers of leaves these deer survived the destruction of eastern forests after the arrival of European settlers.

Badger sett. The number of badgers living in a sett varies between five and twelve. From the entrance a short, straight tunnel leads to a number of other tunnels and chambers, some for breeding and some for sleeping.

although in Britain it lives also on moors. Roe deer and fallow deer are the other native deer of Europe, and their equivalent in North America is the white-tailed deer.

The unique feature of the deer family is the antlers sported by the male. Musk deer and water deer are the only species without antlers, and in reindeer and caribou both sexes have antlers. Unlike the horns of cattle and antelopes, which last for a lifetime, a deer's antlers are shed and regrown every year. They are made of bone and, while growing, are covered with skin – the velvet, but this later withers and falls off. Antlers are used as weapons and as a display of status. Why they should be shed every year is a mystery.

The European badger is a social carnivore that lives in small groups, called clans. A clan may consist of only one boar and one sow, or of up to a dozen badgers. The clan has a territory, the boundary of which the badgers mark with scent. Within the territory there is a system of paths linking the feeding grounds and drinking places with the burrows, or setts. A sett may have several entrances and a labyrinth of passages with sleeping chambers. Although badgers are carnivores, they eat a variety of foods. Earthworms are a favourite and they kill small birds and mammals, but they also eat fruit.

Centre: **the wild boar is the progenitor of the domestic pig, and the most feared inhabitant of deciduous forests. Its tusks can inflict savage wounds.**

The badger of Europe and Asia is a carnivore by pedigree but an omnivore by habit. Its favourite food is earthworms while small mammals and bulbs are also in the diet.

Rodent swarms

In terms of number of species, rodents are the most successful group of mammals alive today. There are about 1800 species in three major groups. The sciuromorphs, or squirrel-like rodents, include squirrels, beaver, kangaroo mice, pocket gophers and the mountain beaver or sewellel; the hystricomorphs, or porcupine-like rodents, include the porcupines, guinea pig, cavies, coypu and cane rats; and the myomorphs, or mouse-like rodents, include the rats and mice, jerboas, dormice, voles, hamsters and mole rats.

There are very few places not inhabited by rodents; even Antarctic islands have been colonised by house mice and common rats. The rodents have taken up almost every way of life, although none is marine and the flying squirrels do not fly properly. The mole rats and pocket gophers live underground, the squirrels and many others are climbers, water voles and beavers are aquatic, lemmings live on the Arctic tundra, jerboas in the driest deserts and mountain voles in the high Himalayas.

One reason for the rodents' success is their teeth. The incisors are self-sharpening chisels. Only the front surface is coated with hard enamel, so the back wears more rapidly and a sharp edge is maintained as the rodent gnaws hard materials. Because growth is continuous the incisors are replaced as quickly as they are worn away. There are no canines and a gap called the diastema separates the incisors from the cheek-teeth. The rodent can pull its lips into the diastema to seal off its mouth so that it can gnaw a nut

American porcupines spend much of their time in trees, sleeping during the day and feeding on leaves and bark at night. Their quills are not so long as those of the African porcupines and are hidden beneath the fur when the animal is relaxed.

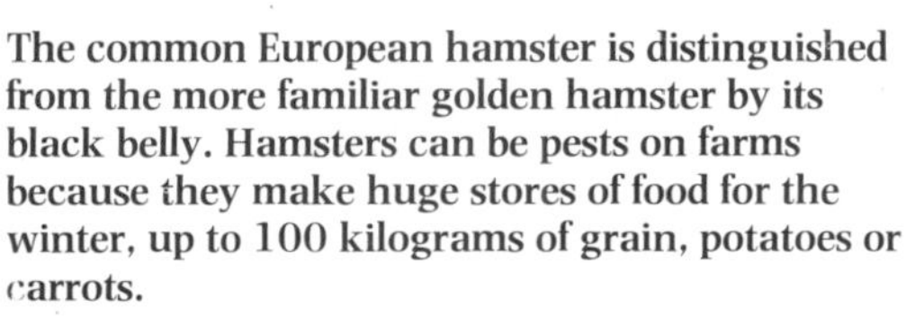

The common European hamster is distinguished from the more familiar golden hamster by its black belly. Hamsters can be pests on farms because they make huge stores of food for the winter, up to 100 kilograms of grain, potatoes or carrots.

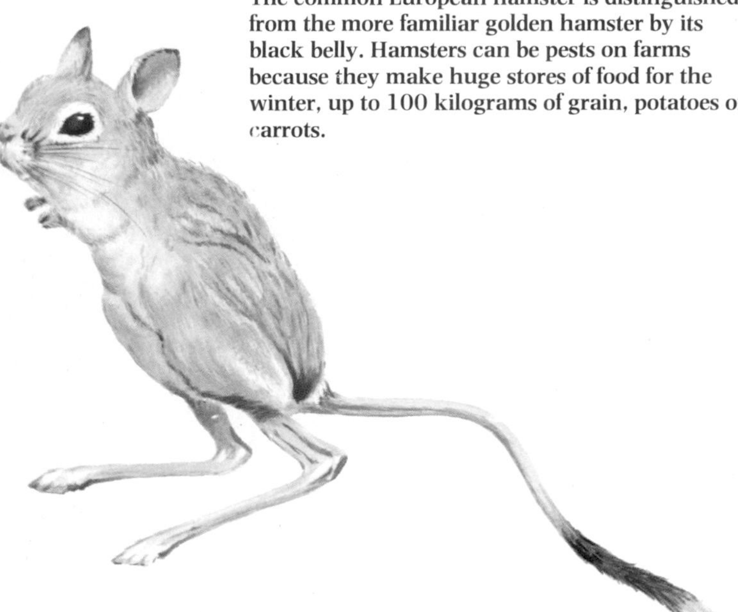

The jerboa lives in the deserts of Africa and Asia. It looks and behaves like a miniature kangaroo. It hops on its hindlegs and uses the long tail for balance.

The story of the lemmings' mass suicide is a myth. As with many small mammals, there is a cycle of abundance. When Norway lemmings become very abundant they spread out from their mountain homes. This migration is most spectacular when they are channelled down valleys, and many die of starvation, predation or drowning as they cross rivers.

shell or excavate a tunnel with its incisors without getting bits in its mouth.

A rapid breeding rate has also contributed to the rodents' success. Many species have large litters and often several a year, so although they are the main prey for many predators, they can recover their numbers quickly. Rapid breeding has enabled rats and mice to overrun islands when they have got ashore accidentally from ships, and it makes them virtually impossible to eradicate.

In view of the numbers and feeding habits it is not surprising that many rodents have become pests – through eating either growing crops or stored food. Despite huge expenditure of money, effort and expertise rodent control has not been able to do more than keep down numbers, except on a very limited scale. If rats could be exterminated, problems of food shortage in some countries would be virtually solved.

Part of the large family of rodents (clockwise from top left): fat dormouse; brown rat; common dormouse; African mole rat; water vole; and birch mouse.

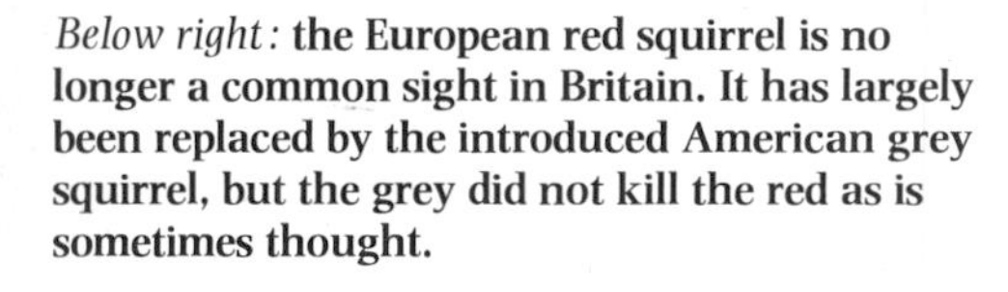

Below right: **the European red squirrel is no longer a common sight in Britain. It has largely been replaced by the introduced American grey squirrel, but the grey did not kill the red as is sometimes thought.**

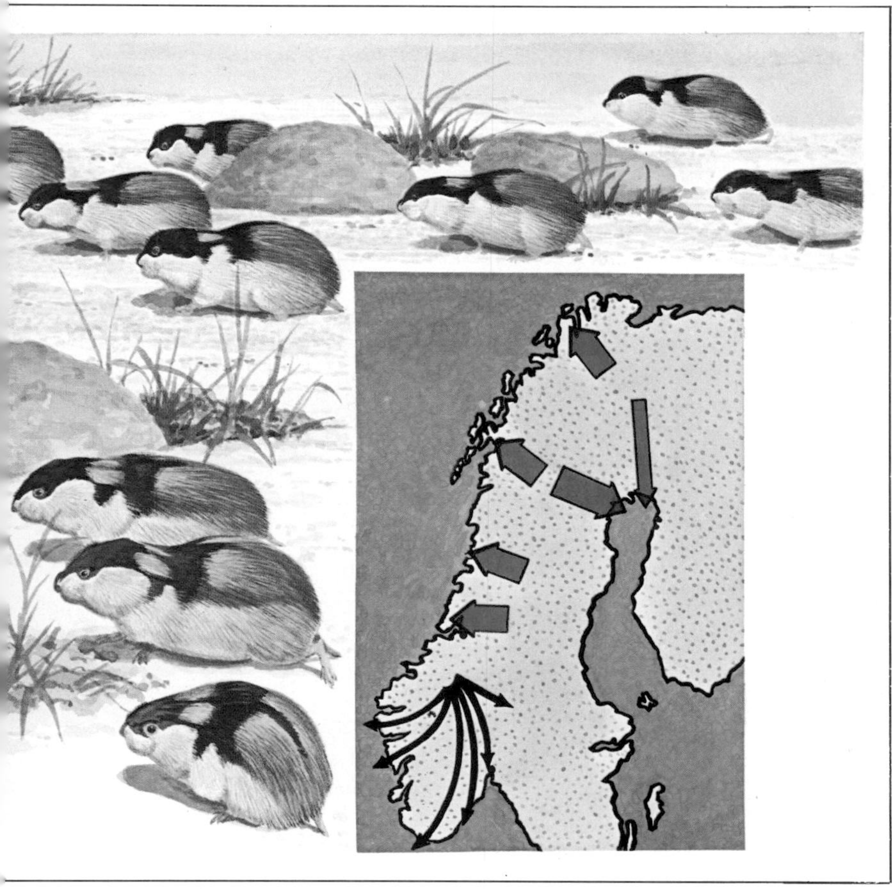

Rabbits and hares

The rabbits and hares, along with their relatives the pikas, used to be classed as rodents because they share the characteristic of continuously growing incisor teeth used for gnawing and a diastema, the gap between the incisors and cheek-teeth. However, they possess two pairs of upper incisors, compared with the rodents' single pair, and they are now known to be no more closely related to rodents than are any other mammals. They have been classed, therefore, in a separate order, Lagomorpha.

Common names of rabbits and hares are rather confusing because the jack rabbits are hares, and the varying hare is also called the snowshoe rabbit. Rabbits and hares are soft-furred animals with long ears, long hindlegs and short tails. They feed on grass, leaves and bark, so they become pests on farms and in forestry plantations. To cope with the large amount of cellulose in the diet and with the often poor quality of their food, the lagomorphs have a specialised digestive process that parallels the many chambered stomach of the ruminants (p. 51). Cellulose is digested by micro-organisms in the caecum, a side branch of the intestine, and food is passed through the digestive system twice by the animal eating its droppings. After the first passage soft droppings are produced, and after being eaten again and redigested they

A brown hare at rest, but still alert. Its ears are ready to pick up the slightest sound and its bulging eyes give all-round vision.

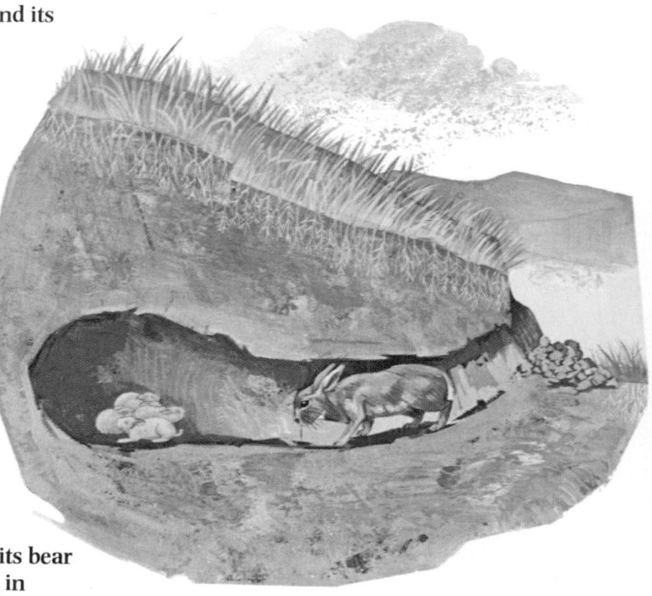

Above: **rabbits bear their young in burrows, where they stay until they are well grown.**

Above: **pikas, or mountain hares, live at high altitudes in mountains where they spend the winter under the snow living on stored food.**

Right: **hares are born on the surface and they soon leave the nest, or form.**

Right: **the black-tailed or Californian jack rabbit lives in the west of the United States. It can be found in deserts where it takes advantage of any shade and uses its huge ears as radiators.**

Below: **by contrast, the snowshoe rabbit has short ears, which helps to conserve body heat in cold weather.**

Bottom: **the desert cottontail, one of several species of cottontails which are found all over North America and parts of South America. These rabbit-like animals are so-called because of the fluffy appearance of their tails.**

are excreted a second time, as hard pellets. This is called refection and it enhances the assimilation of Vitamin B, protein and other substances.

Lagomorphs are absent from Madagascar, but they have been introduced to southern South America, Australia and New Zealand, where they have become pests. They live mainly in open country and rely, for protection, on speed, together with sharp hearing and all-round vision from the bulging, laterally-placed eyes.

The jack rabbits of the North American deserts live in conditions of extreme heat. They have very large ears which are used as radiators to help keep the body cool. The Arctic hare of the tundra, on the other hand, has short ears to reduce heat loss. Both the Arctic hare and the varying hare are white, at least partly for camouflage in the snow. However, the Arctic hare remains white in summer after the snow has melted. The mountain hare, never turns white in Ireland but is white for half the year in eastern Europe. The forest rabbit is unusual in that it is confined to Amazon forests, and the swamp rabbit and marsh rabbit live in marshes and swim regularly.

The pikas are smaller than rabbits and hares and have short, round ears and legs of equal length. Some live at up to 6000 metres in the Rocky Mountains and the Himalayas; others live in the steppes. They store grass, twigs and pine cones under rocky ledges for winter use, after drying them in the sun.

Hares give birth to their young in a nest on the surface of the ground. The young are well developed, with a coat of fur and their eyes open. They are soon active. Rabbits give birth to helpless, naked young in a burrow. When conditions are bad, through overcrowding or food shortage, lagomorph mothers re-absorb some or all of the embryos back into the body before birth as a means of population control.

Hibernation

In winter mammals face periods of low temperatures and shortages of food. A storm with rain or snow and wind can drain the energy reserves of a small mammal as it struggles to keep warm. Similarly a deep fall of snow or a heavy frost can make its food supply inaccessible. One solution is to retire to a snug nest and either subsist on stored food or go to sleep. Badgers and squirrels, for instance, lie up during spells of bad weather. Some mammals retire for the whole of the winter in a more permanent sleep, known as hibernation.

Hibernation is more than a deep sleep. The body undergoes many changes in its physiology and some aspects are still not understood. The process is controlled by the hypothalamus in the brain, and the changes start well before the onset of hibernation. Before the weather turns cold and food disappears, the animal starts to store energy as fat. As well as ordinary 'white' fat, it forms 'brown' fat which is particularly energy-rich.

At the onset of hibernation the body temperature drops to the level of the surroundings, but it is still under control and the body is not allowed to freeze. The heart rate also drops, from 190 to 20 beats a minute in the hedgehog, and breathing slows down. The animal becomes unconscious and cold to the touch, and some species can be handled without being aroused. In this state the hibernating animal uses very little energy. It is just 'ticking over'. However, by the end of the winter it will have used up much of its reserves. If an animal does not put on enough fat before hibernating, it may die in its sleep.

Hibernation is not continuous and animals wake up at intervals. They remain in the nest, or they may come out to look for food. Bats wake up and look for new quarters if their roost gets too cold or search for food if the weather is warm enough to bring insects out.

Arousal from hibernation uses up a large amount of energy. The brown fat is burned

The hazel dormouse spends half the year in hibernation, in a sleep so profound that the animal can be taken out of the nest and not be disturbed.

Opposite bottom: **many species of bats hibernate in winter, as these Schreiber's bats are doing, when there is no longer enough insect food to keep them active.**

Above: **this hedgehog is slowly waking up from hibernation, which can take several hours as the hedgehog's body may be as much as 10°C below normal.**

Right: **a hibernaculum. The hedgehog makes a pile of leaves and grasses and then combs them into a ball around him.**

Above: **the golden mantled ground squirrel goes into deep hibernation in autumn, after fattening up on seeds stored in its burrows.**

Above right: **the Alpine marmot is a ground-living, burrowing rodent of the squirrel family. It makes a nest of dry grass in a burrow and the whole family will hibernate together for up to six months.**

to provide an intense warming around the major blood vessels. The blood then carries this heat to the brain and other organs and normal activity starts. Because so much energy is used, too many arousals during winter can drain the reserves and lead to death. Disturbing bat roosts in winter can severely deplete populations.

Some mammals sleep for long periods in the winter without undergoing the major physiological changes associated with true hibernation, and they are said to undergo winter dormancy. The brown bear lays down a store of fat from the autumn harvest of berries, and when the snow begins to settle it retires to its den. Its heart beat drops from 40 to 10 beats a minute and breathing slows, but its temperature drops only a few degrees.

Fresh waters

Nearly all mammals can swim including even bats, but the giraffe is an exception because its centre of gravity is so high that it would topple over. Yet, few mammals have taken to an aquatic life, although many enter shallow water to feed or wallow. Moose, for instance, wade into shallow water to graze aquatic plants, and brown bears catch salmon without any special adaptations for an aquatic life.

A mammal's body is naturally buoyed up by the air in its lungs and by the air trapped in its fur. In a land mammal water eventually penetrates the fur, which becomes water-logged, and the mammal sinks, but aquatic mammals, such as beavers and otters, have fur that is very dense and water repellent. When wet, the outer guard hairs form a mat over the short underfur. When the animals emerge from the water, the guard hairs look spikey and the animals shake themselves to throw off the water. The platypus and water shrew have velvety fur, which, as there are no guard hairs, becomes very wet. However, as they run through their narrow burrows their fur is squeezed dry. Without a burrow, a water shrew remains damp and soon dies of cold.

Most mammals swim by 'dog-paddling', using the same limb movements they would employ on land. The aquatic mammals are aided by having webbed feet, and the tail is often flattened or fringed with hairs to act as a rudder. Strong swimmers, like the beaver, the otters and the otter shrews of Africa, swim by beating the tail up and down and making undulating movements of the body.

Typically, aquatic mammals have their eyes, nostrils and ears on the top of the head so that the animal can be fully aware of its surroundings even when almost completely submerged. This is best seen in the hippopotamus and capybara. The former spends the day in the water and emerges at night to feed on land. The capybara, which looks like a miniature hippo, feeds on the bank and dives in when alarmed. Nostrils and ears are closed underwater, but the desmans of hill torrents are thought to use their sense of smell underwater. A good set of whiskers is an asset when looking for food or avoiding underwater obstacles. Desmans feed on small animals dislodged by strong currents, and they detect them at a range of 100cm by the vibrations in the water.

Above: **the beaver is highly skilled in the damming of streams and the construction of its home, known as a lodge. Here the beaver is felling trees for its dam and for the bark, one of its food sources.**

Below: **the capybara (foreground) is a common and widespread mammal in South America, where it is to be found in woodland close to lakes and swamps. The largest of all living rodents, the capybara is the prey of the jaguar (background), which is equally at home in the water as it is on land.**

Beavers

The beaver is an amazing animal which creates its own aquatic habitat. By damming a river with a mass of logs, sticks, stones and soil, a colony of beavers creates a lake where it can live in safety. A lodge built of the same materials or a burrow in a bank provides a secure home, safe from land-based predators. During the summer the beaver feeds on land, but in the autumn it takes to felling trees and carrying the twigs to the base of the lodge, where they are stored as a winter food supply. Small twigs are carried in a mass, either held under the chin or pushed along, but large sticks are dragged in the mouth. A runway or canal is sometimes excavated to make haulage easier.

Above: **a hippopotamus threatening. It spends the day in water to keep cool and comes ashore to feed at night.**

Below left: **the otters are a widespread group of carnivores, which hunt along the banks of rivers and lakes and sometimes along sea coasts.**

Temperate grasslands

Typical bighorn country in the Yellowstone National Park. This type of grassland gradually fades into the big prairies of the mid-west which makes up the large part of the temperate grassland habitat.

Prairie dogs

Before the prairies were cultivated, they were the site of huge prairie dog 'towns'. One covered an area of Texas 360km long and 160km wide, and contained an estimated 400 million prairie dogs. Within these enormous colonies there is a well ordered social life. The unit of social life is the coterie, a group of prairie dogs which live in a friendly community consisting of an adult male, several females and their offspring. When a coterie gets too big for the territory, some members emigrate and either join other, smaller coteries or start new ones at the edge of the colony.

Prairie dogs are named after the barking call of alarm with which the colony is alerted to danger. Despite these warnings the colonies attract many predators, such as hawks, black-footed ferrets, coyotes, foxes and badgers.

Below left: **the pronghorn of North America is an antelope-like grazing animal which has the reputation of being the fastest land mammal after the cheetah.**

South of the taiga woodlands and where it is too dry for deciduous woodlands there are huge areas of grasslands. They are situated in the interior of the continental landmasses of North America and Eurasia, where they make up the prairies of the United States and Canada and the steppe which spreads from Hungary, across the USSR to China. Similar grasslands are the South American pampas, the South African veldt and the Australian downlands. Typically these regions have cold winters and hot summers.

Rainfall is usually in spring, when the ground may flood. This is followed by rapid growth of grasses which provide first lush grazing and then an abundance of seeds for many animals to eat. With the grass there are plenty of herbs growing from bulbs. Short shrubs such as sagebrush and heather grow in some areas, but trees are mainly confined

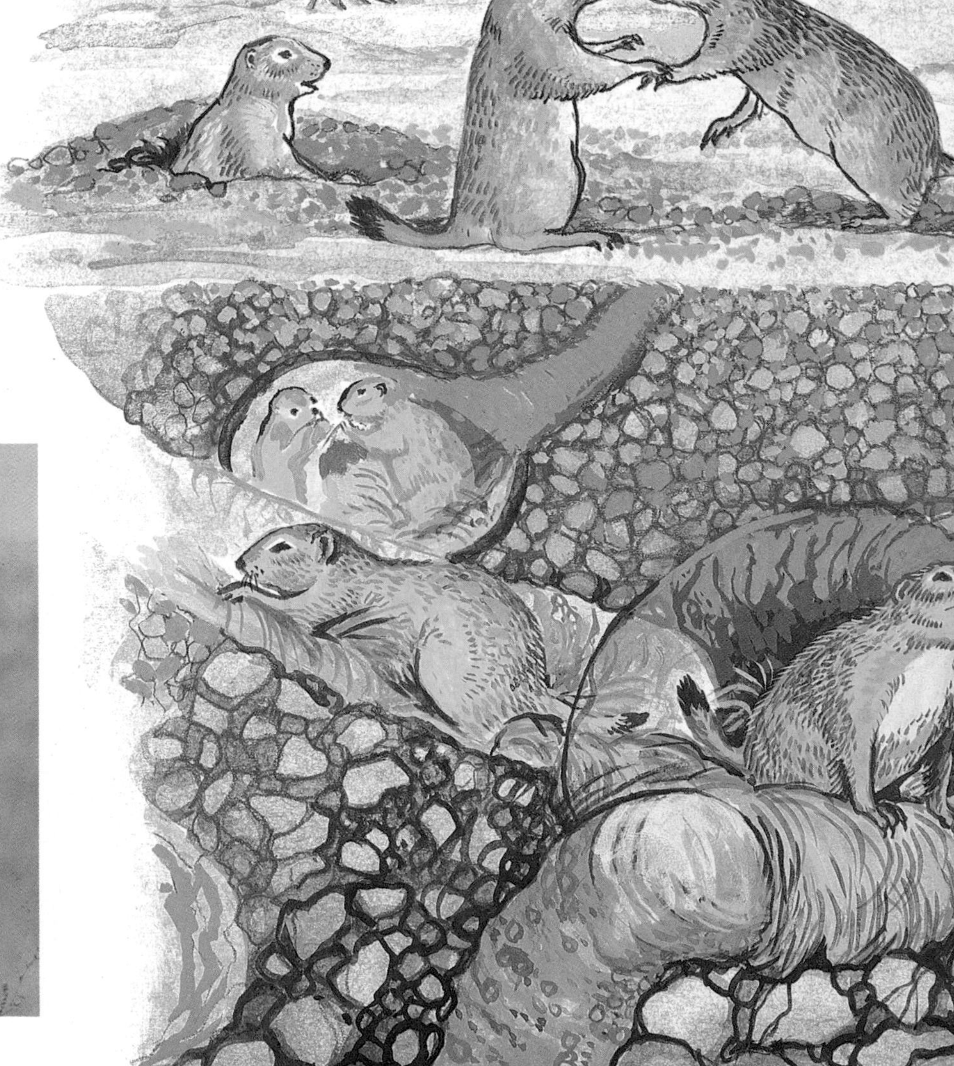

Above left: **bison once covered the Great Plains of North America in huge herds but were nearly wiped out. They now survive on the reserves, which are almost all that are left of the prairies.**

Above right: **prairie dogs keep watch at the burrow entrance. A barking alarm call will send the colony running to safety.**

Below: **section through a prairie dog 'town'. Prairie dogs share their burrows with rattlesnakes, burrowing owls, and black-footed ferrets, which also prey on them.**

to river and lake banks where there is sufficient moisture. Fires sometimes sweep the grassland, so preventing trees from becoming established even in favourable places; fires also reduce dead stems to ash and thus enhance the fertility of the soil.

As in other open habitats, such as tundra and desert, where there is no shelter the large mammals are mainly fast-running animals that live in herds, and the small mammals are burrow-dwellers. Examples of large mammals are wild horses and asses. However, fast-running herds have little defence against mounted men with firearms, so these animals have become rare. Wild horses and asses of the Asian steppes barely survive, and the bison and pronghorn of North America were saved only at the last moment. There were an estimated 60-70 million bison before white settlers reached the prairies. By 1889 there were only 500 left, but their numbers have since increased under protection.

The small mammals of grasslands are mostly rodents, such as the steppe lemming, the pocket gophers, the mole rat, which burrows with its teeth and eats bulbs and roots, and the common and black-bellied hamsters. Rather larger are the ground squirrels such as the marmots and susliks of Eurasia and the prairie dogs of America. These animals have had their ranges greatly reduced by the ploughing up of the grasslands for cereal crops.

Horses and zebras

During the course of evolution, the horse family has changed from the dog-sized *Hyracotherium*, which ran on toes and ate soft leaves, to the familiar horses, asses and zebras which are much larger animals, run on hooves and eat grass. This evolution is associated with a change of habitat from swampy forests to dry plains. From being small animals that slipped through the undergrowth, horses have become fast runners to escape enemies. Their legs became longer, and they raised on tiptoe until they were standing on the nail of a single toe—the hoof. At the same time their teeth developed complex ridges of sharp enamel to deal with tough grass stems.

Much of the evolution of the horse family took place in North America, but then the animal disappeared from there around 10 000 years ago and was only reintroduced in the 16th century, by the Spaniards. Wild horses survived until recently on the Eurasian steppes and in European forests. The forest tarpan became extinct one century ago, as did the steppe tarpan, while the Przewalski horse survives in captivity, and perhaps in the wild in Mongolia. Wild horses live in herds of about a dozen, led by a stallion which protects them from enemies. Horses were first domesticated 2500 years ago and, as mounts for soldiers, they have had a considerable influence on the course of human history.

Wild asses live in deserts and steppes of north-east Africa and of Asia. The African ass is almost extinct as are some of the Asian asses: the kulan and the kiang. The Syrian ass is extinct, but the onager exists in a few populations numbering a few hundreds.

There are three species of zebras: the plains zebra, the mountain zebra and Grèvy's zebra. All live in Africa south of the Sahara and a fourth, the quagga, is now extinct. The species of zebras are identified by the patterns

Right: the Damara zebra is a subspecies of the common zebra and lives in South-west Africa. The stripe pattern is often very irregular.

Below left: a herd of horses in the Camargue of southern France. This is one of many places where horses live in a semi-wild state.

Bottom left: a herd of Grevy's zebras among thorn trees on the African savannah. In this species the males hold large territories through which other zebras wander.

Below: the progressive evolution of the horse family from the small ***Hyracotherium***, or ***Eohippus***, to modern horses. From standing on four toes like a dog, they developed the hoof, which is the nail of a single toe. At the same time their teeth changed to cope with new diets.

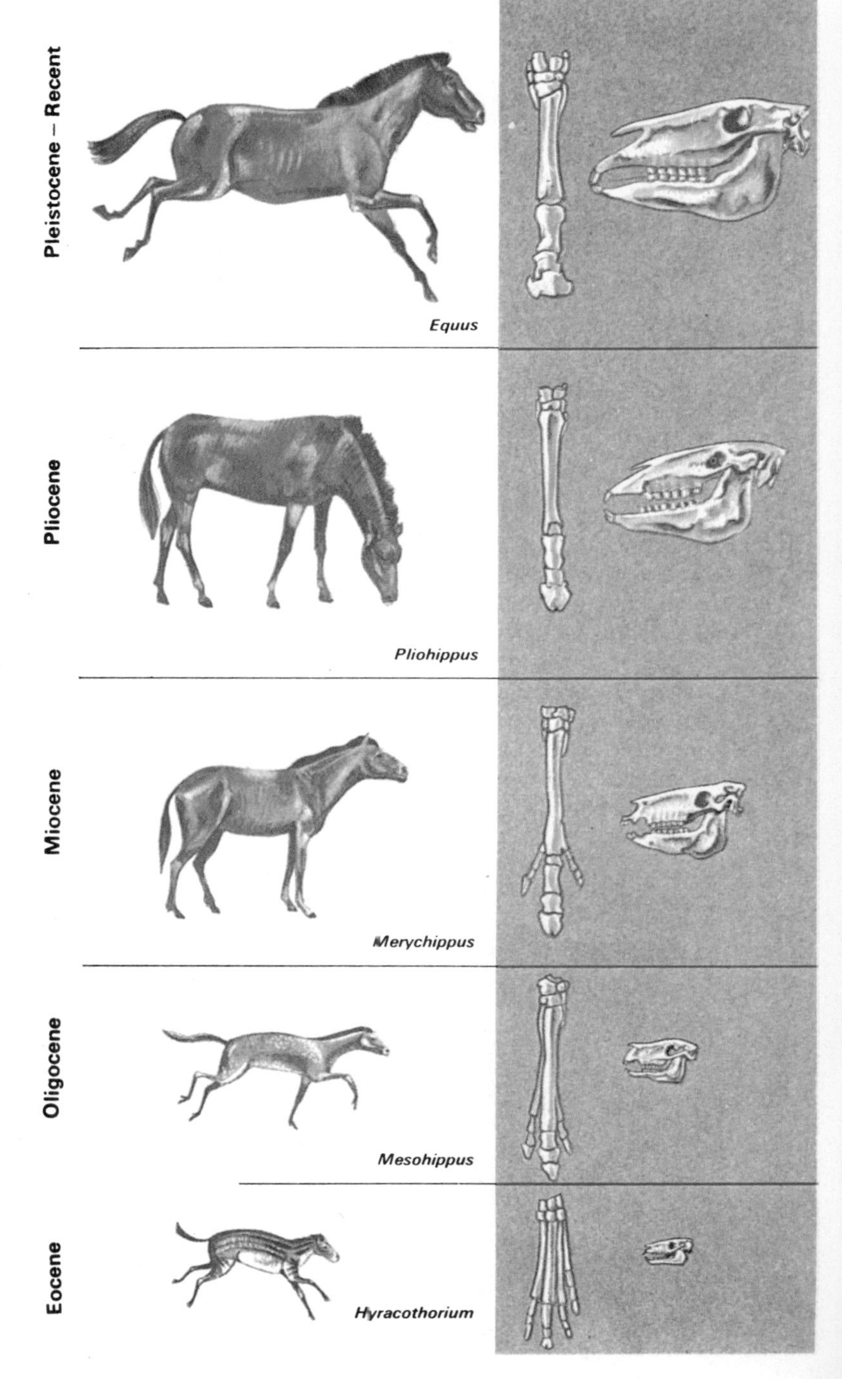

Left: **the onager is a race of the Asian wild ass found in Iran and India.**

Below: **Przewalski's horse is the last truly wild horse. Some may still exist in the wild but there are plenty in captivity.**

of stripes, and each individual zebra has its unique variation of the species pattern, like human fingerprints. Apart from being a means of identification, the function of the stripes is not certain, but in a heat haze zebras are far less conspicuous than other large mammals.

Plains zebras and mountain zebras live in small herds under the leadership of a stallion which fights off other stallions. The herd stays together when fleeing, and the zebras will search for a missing member, but they often mix with other herds at water-holes or when migrating. The young stallions live in separate, bachelor herds. Mountain zebras live in dry country and dig for water in dry river-beds. Grèvy's zebra stallions hold territories of up to 10 sq km, although they will allow other stallions to pass through. They do not control a herd of mares but court individual mares, which they chase into the centre of the territory for mating.

Mountains

Climbing a mountain is rather like travelling towards the polar regions. The temperature steadily drops, at a rate of 1°C for every 160 metres, and the vegetation changes until there is nothing but bare rock and snow. Below the completely barren peaks and above the treeline, there is a zone of sparse vegetation equivalent to the Arctic tundra and called the alpine zone. The atmosphere is very dry, because the air, as it moves up over the lower slopes, releases its moisture, which falls as rain and quickly flows away. The aridity is enhanced by high winds; also, the sun's rays in the thin atmosphere can be very strong and they heat the ground though not the air. Mountain mammals, therefore, have to combat drought and sparse food as well as cold weather. A few mammals are specialist mountain dwellers, but when conditions are favourable they are joined by others.

Most mountain mammals are plant-eaters. The smaller ones, such as the snow vole of Europe, the chinchilla of the Andes and Père David's vole and the bobak marmot of the Himalayas, live in burrows, where they are protected from the worst effects of the weather. These animals lay up stores of food in their burrows for winter use. Larger animals rely on thick, shaggy coats to keep them warm.

There are many kinds of hoofed mammal in the mountain ranges of the world. Europe has the chamois, ibex and the mouflon, North America the Bighorn sheep and the Rocky Mountain goat, Africa the klipspringer, but Central Asia has the widest selection including yak, chiru, bharal, tahr and markhor. In South America the vicuna and guanaco of the camel family are the equivalent high altitude animals. The large mountain grazers are incredibly agile. They run up and down almost sheer slopes and perform great leaps to land on tiny ledges and pinnacles. The klipspringer (the name means rock-jumper in Afrikaans) stands on the tips of its hooves and is famous for sure-footed leaping.

The snow leopard and the Tibetan weasel are regular inhabitants of the alpine zone, but they are joined by other carnivores such as wolves, foxes and wolverines. Leopards hunt rodents at 4500 metres and the body of a leopard was found at 6000 metres on Mt. Kilimanjaro. Bears have been seen at great heights in the Himalayas, and their footprints have started stories of the 'abominable snowman'.

One problem peculiar to life at high altitudes is a shortage of oxygen. Atmospheric pressure is low and there is difficulty in getting enough oxygen to the tissues. A mouse taken to 6000 metres becomes comatose and hardly able to crawl, and the effect of 'mountain sickness' is well known. Mountaineers find sustained effort very difficult, although acclimatisation helps through the production of extra red blood corpuscles for carrying oxygen. Mountain animals, such as llamas, have very large numbers of small corpuscles which carry large quantities of oxygen, and their haemoglobin is adapted to pick up oxygen at low pressure.

Left: on the exposed mountain slopes, soil and water are only briefly retained and the vegetation covering is therefore thin.

Below left: the chinchilla, once hunted for its fur, lives in the Andes, from sea level to over 6 000 metres. Chinchillas living at the highest altitudes have small ears and short tails to cut down on heat loss. They are mainly nocturnal, but will bask in morning and evening sun.

Above: the Rocky Mountain goat of North America is not a true goat. It lives above the treeline, where its broad hooves, which have a hard rim and soft inner bed, give it a sure grip on steep, slippery slopes and cliffs.

Above right: the ibex lives in the Alps and surrounding mountain ranges. It is wonderfully agile and can stand on a pinnacle just big enough to take its four hooves.

Centre: the yak of the Himalayas is protected from the cold by its long, shaggy coat. It forms the basis of Tibetan rural economy.

Bottom right: the llama, of the High Andes of Peru, is well-adapted for mountain life. The haemoglobin in its blood is better at picking up oxygen than that in low altitude animals.

Below: the snow leopard of the Himalayan region has long, pale fur and regularly lives on high mountain slopes, where it hunts pikas, ground squirrels, birds and sometimes larger animals like gazelles and goats.

Deserts

A desert has less than 25cm of rainfall a year. Some desert regions may receive this amount every year, but in some places rainfall is sporadic or even non-existent for several years. Desert species are adapted to either surviving long periods without water or making the best use of what is available. Deserts are usually hot, so animals have to avoid or survive extremes of temperature as well as drought; but there are also cold deserts. The polar regions are deserts because they receive very little snow. The Gobi (in Mongolia) and Atacama (in Chile) deserts are also cold deserts.

Desert mammals get their water in two ways: they can drink or they can get sufficient water from their food. Flesh-eating mammals, such as foxes feeding on rodents or rodents feeding on insects, get most of their water from the body fluids of their prey. Plant-eaters get some water from their food, but desert plants are very often dry, and the mammals feeding on them must therefore conserve their water supplies. They do this in several ways.

Water is lost as urine, moisture in the breath, or sweat. Desert mammals excrete a very concentrated urine, being able to flush the urea out of their bodies with far less water than other mammals. Some desert rodents save on the amount of moisture exhaled in the breath, and they do not sweat. They live in burrows during the day, so they avoid the worst of the heat, because the temperature is always much cooler a few centimetres underground, and the air in the burrows is very moist, so less water is lost from the lungs when breathing. The American kangaroo rat and the jerboa of the Sahara are so efficient at saving water that they can live on dry food without any water to drink. Ground squirrels, however, avoid the hottest months by aestivating (staying dormant during the summer).

Larger mammals cannot avoid the desert sun by burrowing. They make the most of the smallest scrap of shade and a thick coat of hair helps to protect them from the sun. Nevertheless they have to sweat to keep cool, but the camel reduces the amount it needs to sweat by letting its body temperature rise during the day. It does not sweat until its temperature has risen to nearly 41°C.

Sweating, breathing and urinating gradually use up the water in the body, but camels, wild asses and other species to a lesser extent, survive dehydration better than other non-desert mammals. A camel can lose 30 per cent of its body weight, compared with 12 per cent for man, and it can quickly replace this by drinking over 100 litres of water in a very short time.

As well as being short of water, desert mammals face a shortage of food. Rodents store seeds in their burrows, but larger animals may have to search for food over wide areas. Camels and antelopes, such as oryx and addax, are great travellers and the hopping gait of kangaroos, jerboas, gerbils, kangaroo rats and kangaroo mice is a method of long distance travel that economises on energy expenditure.

Opposite top left: **even sandy deserts such as the Sahara, in places seemingly unable to support any form of life, are inhabited by some specially adapted mammals.**

The kangaroo rat of North American deserts *(opposite centre)* **and the Cape ground squirrel of Africa** *(opposite top right)* **avoid the heat of the day by sleeping in their burrows and come out at night.**

Opposite bottom: **the camel is the most famous desert animal, renowned for its ability to go for long periods without water. Its physiological mechanisms for survival in hot weather without drinking are a marvel of adaptation.**

Above right: **the caracal or desert lynx of North Africa and Asia is a long-legged cat that frequents deserts. The blood of its prey gives it enough fluid so it need not drink frequently.**

Right: **a scene in a desert of North America. The kangaroo rat is a desert specialist, but the cacomistle, a relative of the raccoon, the fox and the rabbit live in a variety of habitats and are adaptable enough to survive in deserts.**

Savannahs

The name savannah comes from the Spanish word for a treeless plain, *zavana* (now *sabana*); and it is used to describe vast expanses of grasslands with scattered trees where there are long spells of dry weather broken by rainy seasons. The most famous savannahs are those in Africa, especially in East Africa, but the savannah sweeps round the continent in a huge arc that separates the wet forests of West Africa from the dry deserts in the north and south. Other savannahs are found in the drier, monsoon-swept parts of India and to the north and south of the Amazon forests of South America.

Depending on the region, there are one or two rainy seasons which are often very predictable. When the rains come there is a burst of new plant growth and insects emerge. Mammals and birds have a feast after the months of drought, and they breed at this time so that the young will have plenty of food. After the rains have ended, the vegetation dries up, the grasses wither into straw, and the trees and shrubs shed their leaves. Life becomes difficult for the animals.

Fires are common on the savannah and are an important part of its ecology. The flames sweep quickly across the ground, killing young trees and preventing them from smothering the grasses, and the dead grasses are burnt off, leaving their ashes as fertiliser. Fresh growth soon appears from the underground parts of the plants which were not harmed by the fire.

Some savannah species, like giraffes, kudus and elands, can go for long periods without water, but elephants, buffaloes and zebras need frequent visits to water-holes or rivers. Elephants and mountain zebras dig into dry river-beds to find water, and other animals take advantage of these 'wells'. Small mammals, like gerbils and mole rats, store seeds and remain in their cool burrows. Other animals migrate away from the dry areas. The best known and most spectacular migration is that of the wildebeest, a large antelope, round the Serengeti plains of Tanzania. During the wet season the wildebeests spread out to feed on the fresh grass, but as the ground dries out they are forced to move. Huge herds walk in file across country and wear deep tracks in the dry soil. They make their way toward the permanent rivers and wetter country to the west and north of the plains. When the rains start again the

herds return to the plains, having covered 1500km in the year.

The savannahs of Africa support more than 40 species of large plant-eating mammals, as well as swarms of rodents. Until reduced by man, they lived in huge numbers, which was made possible by each species having its own preferred diet or living space, thus avoiding competition. For instance, oryxes live in the driest places, sitatungas are found in swamps, zebra and gazelles prefer open grassland and dik-diks and duikers the wooded areas. Giraffes feed on the highest level, while black rhinos eat leaves at the base of the trees. Even in open grassland the animals do not compete because some select only the choice parts of a grass plant, while others can survive on coarse stems. On the Serengeti plains there is a grazing succession. First come the zebras, which eat large quantities of coarse grass. Then the wildebeests crop the greener parts exposed by the zebras, and topi dip their pointed muzzles between the long stems to get at the leaves. Finally Thomson's gazelles eat the low herbs and the stem bases of the grass.

Above: **African buffaloes are the largest of the hoofed animals on the savannah. They can also be the most dangerous.**

Above right: **eland are large antelopes which have been domesticated for meat and milk.**

Centre: **impalas are medium-sized antelopes which are common in many parts of Africa, preferring wooded country and never moving far from water.**

Right: **the lion is the largest of the savannah predators and, by hunting in groups, it can overcome even buffaloes.**

Opposite top: **the savannah scene is one of expanses of grass with acacia trees or open woodland, depending on the amount of rainfall and extent of any bush fires.**

Opposite bottom: **elephants help to maintain the open countryside of the savannah by knocking over trees to eat the foliage. Where their numbers are high, they can cause enormous damage.**

Antelopes

The antelopes are a large group of cloven-hoofed mammals belonging to the cattle family Bovidae. They include the gazelles, and although they are not easy to define as a group, they are gracefully built and have upward-curving horns. They range in size from the eland down to the dik-diks, and they are found in Africa and Asia, where some species gather in immense herds in open country (or they did until their numbers were greatly reduced). Antelopes were once shot in huge numbers for sport but more serious has been their slaughter in places where they have competed with domestic animals for food or where they have transmitted disease such as nagana and rinderpest.

The klipspringer is an agile antelope of mountain crags; oryx and addax are desert species; bongo, suni and blue duikers live in dense forests; and waterbuck and sitatunga live in wet places. The greatest variety of antelopes is found on the savannahs. On open grassland there are herds of Thomson's gazelles, wildebeests and topis. Impala and Grant's gazelles are seen on the grasslands, but they are browsers (leaf-eaters) as well as grazers (grass-eaters) and are also found in the bush savannah, which is thick with acacia thorn bushes. They are joined in the bush by kudus, dik-diks and the long-necked gerenuk, which stands on its hindlegs to reach a height of $2\frac{1}{2}$ metres.

Asia is not so well endowed with antelopes as Africa. At one time the blackbuck was abundant in the dry woods and semi-desert regions, where it survives without drinking in the hottest months. It formed the main prey of lions, tigers, leopards and cheetahs, but overhunting and destruction of its grazing grounds has made it rare, and this is one reason for the decline of the big cats. The nilgai, or blue bull, is a larger antelope living in the same habitats, but it will eat leaves from bushes, while the blackbuck eats only grass. The four-horned antelope is unique among hoofed mammals in possessing two pairs of horns.

Below left: **the nilgai (foreground) and the blackbuck, both rare beasts of the Indian plains.**
Below: **the gemsbok of South-west Africa, and** *(bottom right)* **scimitar-horned oryx and addax of North Africa, are desert-dwelling antelopes. They survive long periods without water and feed on sparse, dry vegetation.**
Bottom left: **the klipspringer walks on the tips of its hooves and leaps agilely around rocky crags.**
Opposite centre left: **one of the most striking of antelopes, the sable antelope of African bush country has become rare through overhunting.**
Opposite bottom: **the four chambered stomach of a ruminant. Rumination allows the animal to eat a large quantity of food in a short time and process it efficiently.**

Above: **springboks once lived in huge herds in South Africa. They are named after their ability to make three-metre leaps into the air when alarmed.**

Above right: **the greater kudu has a magnificent set of spiral horns. It is well camouflaged in bush country where it lives in small groups.**

Below: **dik-diks are the dwarves of the antelope world. They live in small parties in dense undergrowth and are named after their alarm calls.**

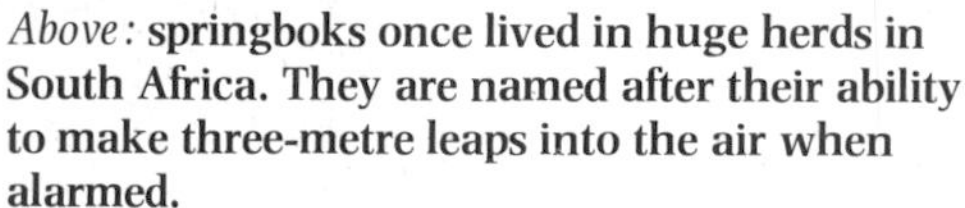

Plant food is not easy to digest because animals do not have digestive enzymes for breaking down the cellulose walls of plant tissue and releasing the nutritious contents. Plant-eating mammals have overcome this problem by employing microscopic protozoans and bacteria to do the job. The antelopes and the other cloven-hoofed mammals, with the exception of the pigs and hippos, digest their food by the process of rumination in a four-chambered stomach. Food is swallowed and stored in the first chamber, the rumen, where cellulose is digested. Then it is regurgitated back to the mouth and chewed again. This is called chewing the cud or rumination. It is swallowed again and passed to the reticulum and omasum, where it is crushed, and then to the abomasum. This, last, chamber is the equivalent of a normal stomach and absorption of the food starts here. One advantage of this system is that the animal can eat large quantities in a short time, then retire to a safe place and process it at leisure.

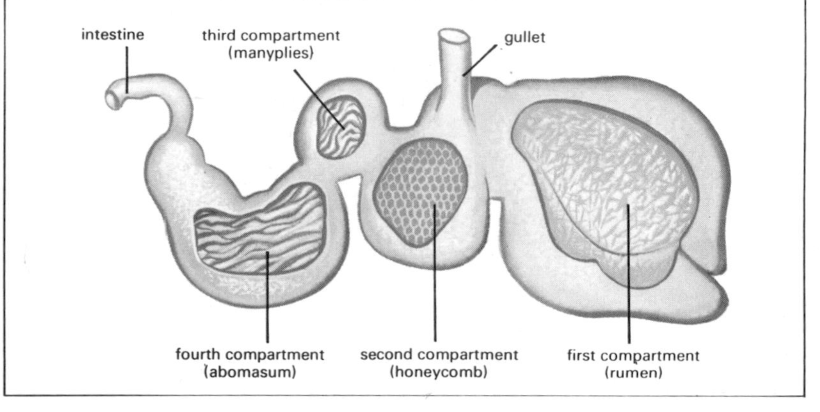

Elephants, rhinoceroses and hippopotamuses

The elephants are the largest living land mammals. The African species is the larger of the two species, with males weighing up to 5300kg; Indian elephants grow to only 4500kg. The African elephant is distinguished also by its larger tusks and ears and a double 'lip' on the end of the trunk.

The great weight of the elephant's body is supported on legs that are built like columns and end in padded soles. Elephants cannot run or jump but they can move at a fast pace. The ears provide a large surface area from which to lose body heat, and the larger ears of the African elephant are due to its living in the open country of the tropics, whereas the Indian elephant lives in the cooler shade of the forest. The tusks are outsize upper incisor teeth which are used for defence and digging for food or water. The trunk is an elongated nose used for gathering food, sucking up water to drink and wash, and for caressing other elephants.

Elephants live in extended families, consisting of an old female, her eldest daughter and their offspring. An elephant calf is suckled for at least four years and is protected by its mother for a few more years. Young bulls leave the family when mature and live alone or in bull herds, but cows stay with the mother.

The food of elephants is mainly grass, with some leaves, fruit and bark, and elephants can do enormous damage by knocking over trees to get at the leaves. Over 200kg of food are needed each day. Only a short length of the teeth surface is in use at a time. As the front portion is worn down, the part behind comes into use. When all the teeth are worn, the elephant dies, at a maximum age of around 70 years.

The five species of rhinoceros belong to the odd-toed hoofed mammal group, the Perissodactyla. They have three hoofed toes on each foot. All five species are now rare, and the Javan rhinoceros is down to a few dozen individuals. The Javan and Indian rhinoceroses have a single horn, made of hairy tissue, and the Sumatran, black and white rhinos have two. The Javan and Indian rhinoceroses also have deep folds in the skin. Rhinoceroses are rather solitary animals, although small groups may be found where they are still numerous, and a calf may stay with the mother after the birth of the next calf. The African rhinos live in dry country but must have drinking water every few days. The black rhinoceros feeds on bushes and scrubs, using its long upper lip to gather twigs. The white rhinoceros has a broad flat upper lip and grazes on grass. (The name 'white' comes from the Afrikaans for 'wide'.) The three Asian rhinoceroses prefer wet swampy country.

The hippopotamuses differ from the elephants and rhinoceroses in having thin skin. The common hippopotamus and the pigmy hippopotamus of West Africa are cloven-hoofed mammals distantly related to pigs. Hippos spend most of the day in water and come onto land to feed at night. They need to keep cool and they are said to 'sweat blood', but the 'blood' is a pink secretion which guards the skin against the sun's rays. Hippos swim well, and by adjusting their buoyancy they can walk on the river bed.

Above left: **the huge ears of elephants may be used to make them look more fierce or to wave away flies. The African elephant, shown here, has an arched forehead in contrast to the concave shape of that of the Indian species.**

Left: **Indian elephants: elephants spend much of their time bathing and use their trunks to squirt water over their thick, yet sensitive skin.**

Opposite top left: **the two African rhinoceroses are easily distinguished. The white rhinoceros (above) has a flat upper lip and the black rhinoceros has a pointed upper lip.**

Opposite top right: **Like all five rhinoceros species, the Indian species is threatened with extinction.**

Opposite centre: **the black rhinoceros was common in parts of Africa until it was hunted for its horn. Rhinoceroses are now mainly found in reserves.**

Right: **the pigmy hippopotamus, of West Africa, is rarely seen because it lives in the forests.**

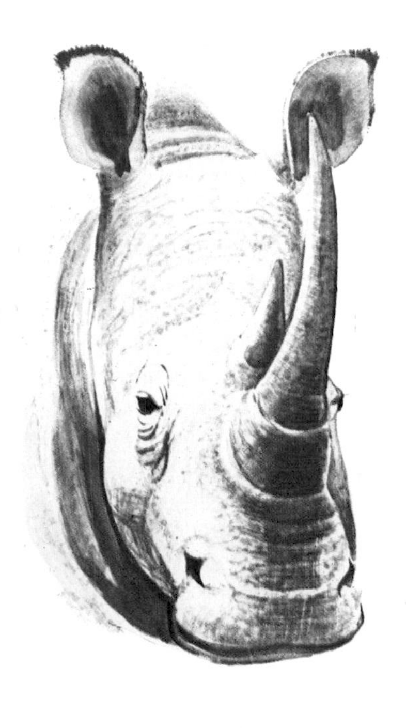

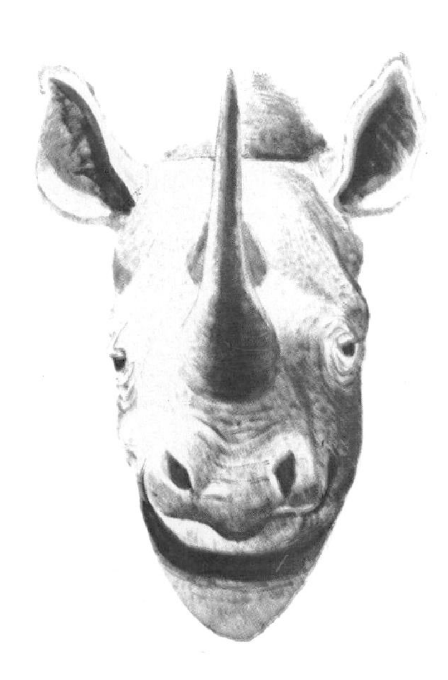

The Carnivores

Although carnivore can mean any flesh-eating animal, it is used especially for members of the mammalian order Carnivora. There are seven families: dogs, bears, raccoons, weasels, mongooses, hyaenas and cats. Not all are predatory animals: most bears and badgers are omnivores, that is, they eat plants and small animals, and the giant panda eats plant food almost exclusively. On the other hand there are animals like baboons which are not strictly carnivores but sometimes kill and eat prey.

To succeed as flesh-eaters, these animals must be well armed to overpower their prey. Their feet cannot be so well adapted for running or swimming that they cannot be used also for striking and handling their prey, and the teeth and jaws must be both weapons and food processors. Carnivores must also be intelligent if they are to find and outwit their prey.

The cats are the most highly developed of the carnivores. They specialise in catching their prey in a short dash, having either lain in wait or approached cautiously within range. Unlike nearly all other carnivores, cats retract their claws into sheaths to keep them sharp for grasping and tearing their prey, or for tree-climbing. The carnivores have well-developed canine teeth, or fangs, for stabbing and holding prey. The cheek-teeth are clearly different from the flat-topped, ridged teeth that plant-eaters use for crushing tough food. The carnivores have sharp edged teeth with points for chopping meat or cracking bones. Two teeth in each jaw, called carnassials, work like scissor blades, for slicing.

Carnivores usually hunt prey smaller than themselves, but even so they must be careful if they are to avoid being injured as the victim lashes out to defend itself. Those that attack larger prey usually hunt in groups, for example, wolves, lions and hyaenas, or are specialist killers like the stoat, which attacks rabbits with an accurate and fatal bite to the neck and can hang on without dislocating its jaw.

The flesh-eating, or carnivorous, animals are at the end of the food chain. Energy from the sun is trapped by green plants in the process of photosynthesis. Plants are called primary producers and they are eaten by animals called primary consumers. The plant-eating animals are, in turn, eaten by the secondary consumers. These are the carnivores, which can be tiny shrews eating beetles or the big cats hunting antelopes or deer. A food chain is an oversimplification, and is only part of an incredibly complex food web with links connecting all the plant-eating animals and their predators. The chart shows a few of the links in the food web of the African savannah. It is complicated by large carnivores, like the lion, eating small carnivores like mongooses or aardvarks, and even a lion sometimes eats beetles.

Red fox
The red fox is an opportunist feeder and is an extremely successful animal that flourishes in towns and survives in the country despite centuries of persecution. Its main prey is rabbits, voles and other small mammals, but many birds also are caught, and the list of food recorded includes weasels, slugs and snails, earthworms, frogs, moths, birds' eggs and a surprising amount of fruit. Hunting tactics include stalking, headlong rushes to catch prey by surprise, or merely stumbling on something edible. Mice and voles which can be heard but not seen in the grass are caught with the 'mouse jump'.

Mongooses *(opposite top)* and polecats *(opposite centre)* are rather similar short-legged, long-bodied carnivores, but they belong to different families. Mongooses, with the civets and genets are viverrids; polecats with weasels, otters, martens and badgers are mustelids.

Above: a cheetah lies in wait for its prey. When a victim is in range, the cat pursues with a rapid sprint.

Above right: Geoffroy's cat of South America uses an ambush followed by a pounce to catch its prey.

The skull of the sabre-toothed tiger *(right)* compared to that of a modern cat *(centre right)* shows how modifications have occurred to the striking and biting elements of the jaw.

Tigers attack large prey, such as this ox, on the move but they often approach smaller animals stealthily. They use their long sharp canine teeth to kill large victims.

Hunters and hunted

Predators and their prey are engaged in a form of arms race. The predator is, through the evolutionary process of natural selection, improving its attack capability. It needs to find victims, and to attack and kill them as efficiently as possible. If it cannot find prey or is unable to make contact and overpower it sufficiently frequently, it will starve and weaken, and then its chances of making a kill will rapidly decline. The prey must maintain its early warning and defence systems. Acute vision, hearing and smell are needed to warn it of danger in time for defensive manoeuvres. A typical prey animal, like a rabbit or antelope, has bulging eyes on the side of its head for all-round vision, whereas predatory cats and dogs have forward-looking eyes with good stereoscopic vision for judging distances in an attack.

The best form of defence is to run away, and prey mammals usually stop and defend themselves only when cornered. Musk oxen run from wolves, and only when overtaken do they stop and form their defensive huddle or *karre*. Strangely, prey mammals sometimes do not defend themselves and allow themselves to be killed without a struggle. They go into a state of shock, and in this state they do not seem to feel pain.

As well as the contest between individual animals, there is an interaction between predators and prey at the population level, which, even after years of study by zoologists, is not fully understood. At one time it was thought that predators kept the populations of their prey under check. So lions were killed in game reserves with the intention of preserving antelopes and other plant-eaters. The actual situation is the reverse: predator numbers are controlled by the numbers of prey (see page 27). When the prey becomes scarce, the predators do not breed successfully and may starve. The key factor in the regulation of animal numbers is food rather than predation, so the prey animals are controlled by the amount of plant food available.

Predators kill only a small percentage of the prey population each year, perhaps around 10 per cent. Most of these will be young animals or old, sick and wounded adults. The depredations will have no effect on the total numbers except in unusual cases. When the numbers of a prey animal drop, most predators can switch to other species. They have their preferences, but these are modified by whatever is most available in terms of numbers and ease of catching.

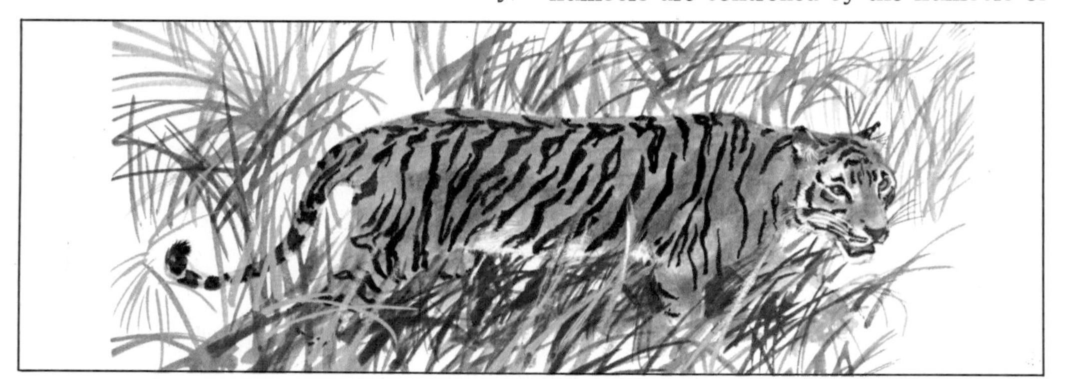

Left: **the tiger's stripes make it difficult to see because they merge with the vegetation and break up the outline of the body. If it had horizontal stripes the tiger would be very conspicuous.**

Below: **zebras live in small herds as a defence against predators, but lions also live in a group, and through communal attacks they can hunt zebras.**

Unlike other wild cats, lions hunt together and feed together. The highest ranking animals in the pride feed first and get the tastiest morsels.

Below: leopards frequently hunt from an ambush in a tree, and after a meal they carry the remains of their prey up a tree. This shows the great strength of the cat. It is a necessary habit to prevent other carnivores from stealing the carcase.

Below right: the cheetah is a very specialised hunter. Unlike most cats, it hunts by day and its main prey are gazelles and smaller mammals and birds.

Rain forests

Above: diagram of a rain forest to show its layers. Different animals live in each layer.

The squirrel monkey *(left)* and howler monkey *(below)* live in the forests of South America, where they clamber and jump among the branches. Both have prehensile tails, a feature of monkeys only from the New World.

Tropical rain forests stretch round the world where there is an even temperature and rainfall of around 200cm a year. They occur over much of Central America and northern South America, in a belt across Africa, and in oriental Asia from India through South-East Asia and into northern Australia.

In these forests there are no seasons and very little annual variation in temperature; changes between night and day are greater than the annual variation. These conditions have stimulated an enormous diversity and richness of plant and animal life. Evolution has 'run riot' in the production of bizarre adaptations for feeding and reproduction. One feature of the forests is that there are many species packed into a small area, perhaps scores of tree species in one hectare, but none are very common. This contrasts with other parts of the world where there are usually relatively few species but abundant numbers of each.

At first, a tropical forest seems a confusing mass of greenery, but it has a definite

structure. There are four layers of trees, starting with the emergent trees; they form a top storey of scattered crowns standing above the rest of the forest. The canopy of continuous foliage of densely packed trees makes a more continuous layer beneath them. Below the canopy there is a layer of medium trees, and then one of small trees and shrubs. Under the tree layer on the forest floor are the low-growing plants. The canopy shuts out much of the light and there is intense competition among the plants to reach it. Trees can make little growth until a neighbour dies and lets in the light, but climbers 'cheat' by growing up tree trunks, and there are many epiphytic plants which grow on tree branches without any contact with the ground.

Rain forests are the homes of many mammals that live above ground. There are hundreds of species of insect-eating and fruit bats, as well as the gliding scaly-tails, a sort of flying squirrel; the flying lemur and flying phalangers; there are leapers, like the squirrels and monkeys; and there are the clamberers, such as the sloths, pangolins, tree kangaroos and small cats. The ground is the home of pigs, peccaries, tapirs, deer and antelopes.

As there are no seasons, plants are flowering and fruiting all the year round so there is always a harvest for mammals. However, despite this richness, the rain forests form a very fragile habitat that is easily destroyed. For instance, many plants are pollinated by bats, which work from species to species as they come into flower. If some trees were cut for timber this might create a gap of a few weeks when the bats were short of food. If this caused the bats to die out, the remaining plants dependent on them would also die out.

The tree pangolin is an African species. It has a long prehensile tail that can be wrapped around branches for support as it climbs.

Opposite bottom: the flying lemur lives in the upper layers and glides on 'wings' of skin stretched between its legs. Here the wings are forming a cradle for its young.

Below: clouded leopards live on the ground in Asian forests but they climb into trees to sleep or lie in wait for prey.

Above: this is the Malayan tapir, related to rhinoceroses and horses. It lives in the forests of Asia, and another species lives in South America. Tapirs live on the forest floor, usually near water and they feed on grass and leaves.

Below: peccaries are relatives of the pigs that live in South America. Some species are forest dwellers.

Primates

The primates, the order of mammals to which human beings belong, evolved from primitive mammals. The first primates lived in North America 65 million years ago. They were rat-like animals that scampered on the ground and only later did they take to living in trees. From North America primates migrated to the Old World and South America, and they are now found only in warmer parts of the world, especially in rain forests.

There are four main groups of primates. The prosimians include the lemurs, bush-babies, lorises and tarsiers, and they live in the Old World, especially on the island of Madagascar. There are two groups of monkeys: the New World monkeys and the Old World monkeys. The former have round heads, flat noses and long tails, which are sometimes prehensile and used as extra hands. The latter have longer noses and their tails are sometimes short (a mere stump in the Barbary ape) and never prehensile. The fourth group is the apes, of which there are only four living types: gorilla, chimpanzee, orang-utan and gibbon. The main features of apes are that they have no tail and that their arms are much longer than their legs.

In the course of their evolution into tree-dwelling animals, the primates developed several characters that were to be important to the human species. Early mammals relied largely on smell, but the primates became increasingly reliant on vision. Their eyes became set in the front of their heads with an increase in stereoscopic vision, which is useful for judging distances when jumping. The hands and feet changed from paws to grasping organs with strong, flexible fingers and opposable thumbs. Gripping a branch is similar to the 'power grip' for holding a hammer, and some monkeys also have the 'precision grip' of thumb and index finger for delicate manipulation. These characteristics allow humans to manipulate and examine objects very closely.

Opposite top: **a slow loris.**

Opposite bottom left: **a squirrel monkey suspended by one arm and a prehensile tail.**

Opposite bottom right: **a ruffed lemur, one of a primate family confined to the island of Madagascar.**

Right: **a hoolock gibbon, an ape which swings overarm. This female has an infant clasped to it.**

Below left: **a selection of colourful guenon monkeys from Africa; from top to bottom: the greater white-nosed, the lesser white-nosed, the redtail, and the red-eared.**

Below centre: **monkey hands: marmoset hand (top) with long claws and non-opposable thumb; capuchin (centre) with semi-opposable thumb; and macaque (bottom) with fully opposable thumb.**

Below right: **all primates have the power grip, but only a few have the precision grip for delicate manipulation of objects.**

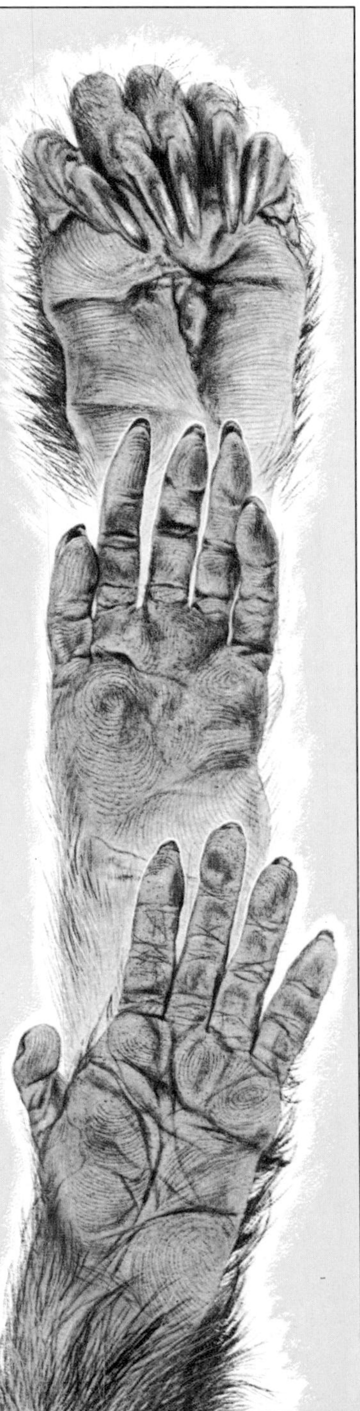

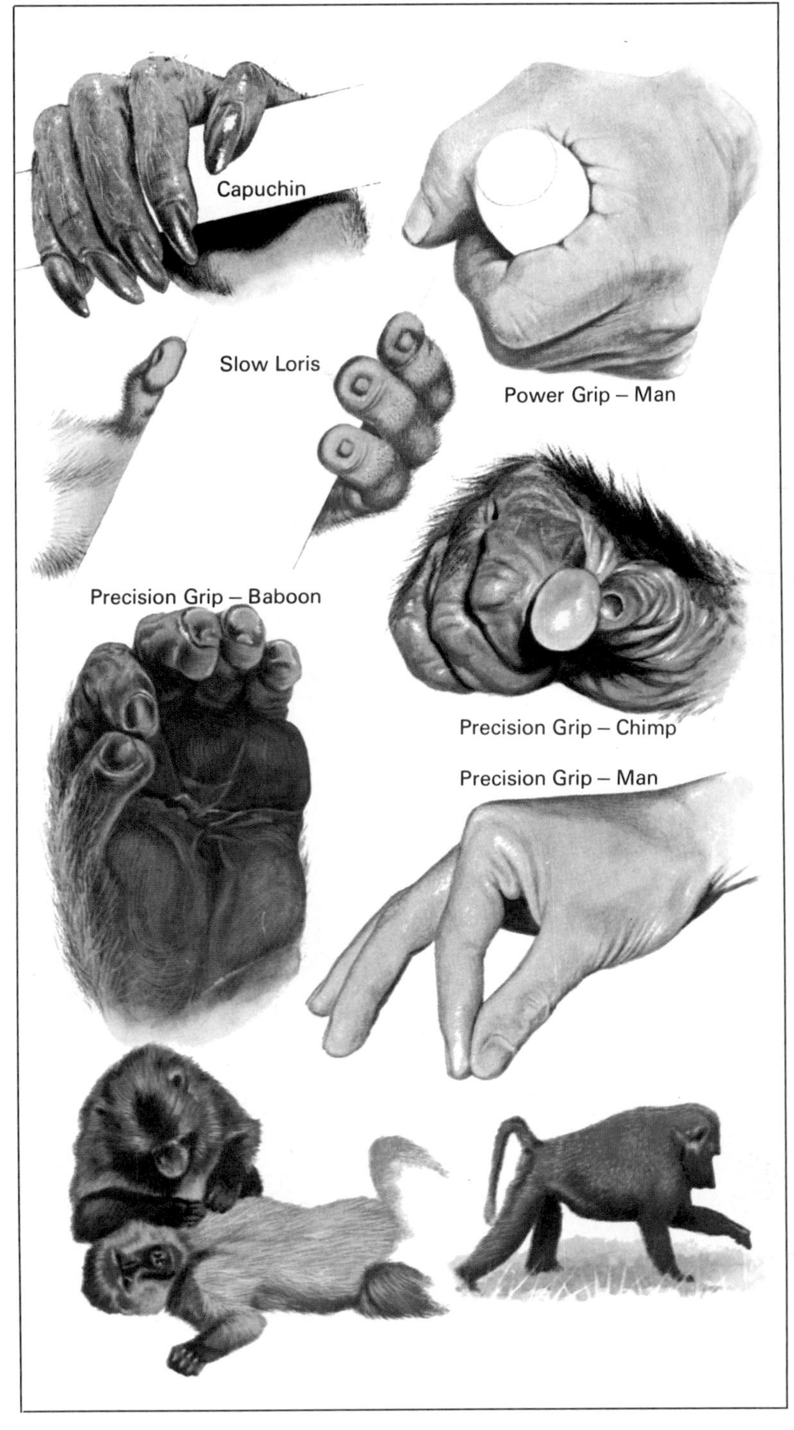

Primates move in three basic ways. Some of the prosimians, such as bush-babies and tarsiers, tree-hop; they have long back legs and jump like frogs, landing back feet first. Most primates walk or run on all fours, both on the ground and along branches, and arms and legs are of equal length. The apes are arm-swingers or brachiators; they have long arms and swing through the trees at a great speed. In addition, a few primates spend some of their time on the ground and occasionally walk upright, both to get a good view of danger and to carry food. Human ancestors must have 'come down from the trees' and learned to walk upright, thereby freeing their hands for carrying tools, weapons and loads. However, the monkey that spends most of its time on the ground is the baboon, and it is strictly quadrupedal.

The primates have also evolved a high degree of intelligence, which may be linked with the problems of living in a three-dimensional world and of having a complex social life. As primates are increasingly studied in the wild, more is being discovered about their abilities to learn. Chimpanzees frequently use tools, such as a twig for fishing termites out of their nest, and Japanese macaques learn new habits from each other. When one member of a troop learned to clean sand from sweet potatoes given to the macaques by scientists, other members learned the trick from it.

Successive stages in a leap of the tree-hopping locomotion of an Indris. The animal is mainly vertical. The line sketch illustrates the relative lengths of the arms and legs. All tree-hoppers have legs that are longer than the arms.

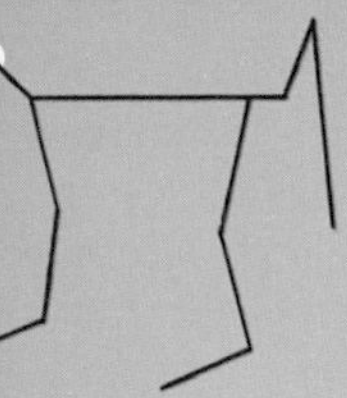

Stages in the gait of a baboon running along the ground. The body is horizontal and the arms and legs are about the same length.

Stages in the arm-swinging of a gibbon. In all the brachiators the arms are very much longer than the legs.

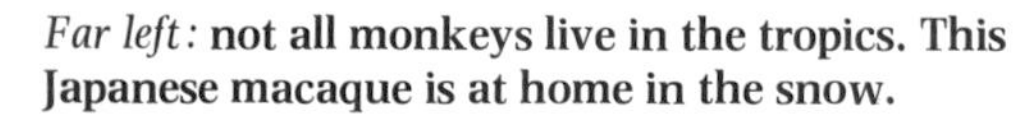

Far left: **not all monkeys live in the tropics. This Japanese macaque is at home in the snow.**

Above: **the three main types of primate locomotion. Some prosimians are tree-hoppers, apes are arm swingers, but most species walk along branches or on the ground.**

Left: **a male mountain gorilla, or 'silverback', is not as fierce as he looks, although he is immensely strong.**

Below: **group defence: a party of baboons on the move with the dominant males guarding the females with young. When they meet a leopard, the males advance to cover the troop's retreat.**

Above right: **chimpanzees are anatomically very close to man. Their brain, teeth and the relative lengths of their limbs make them more like man than are the other anthropoid apes.**

Right: **apes, with the exception of gibbons, build nests to sleep in. This chimpanzee has made a platform of leafy branches.**

Bottom right: **the mandrill male has a spectacular blue muzzle.**

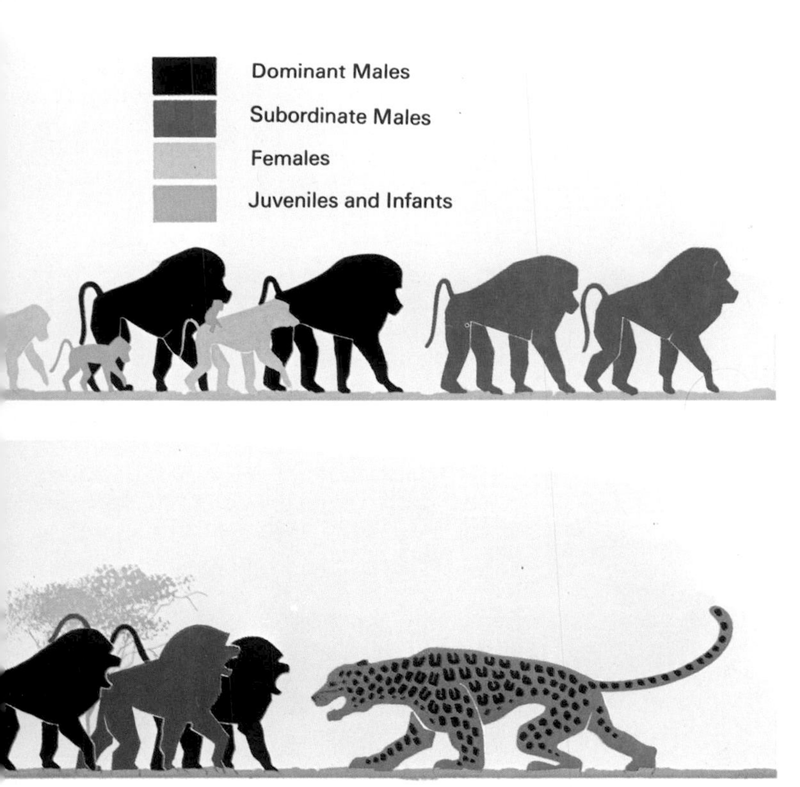

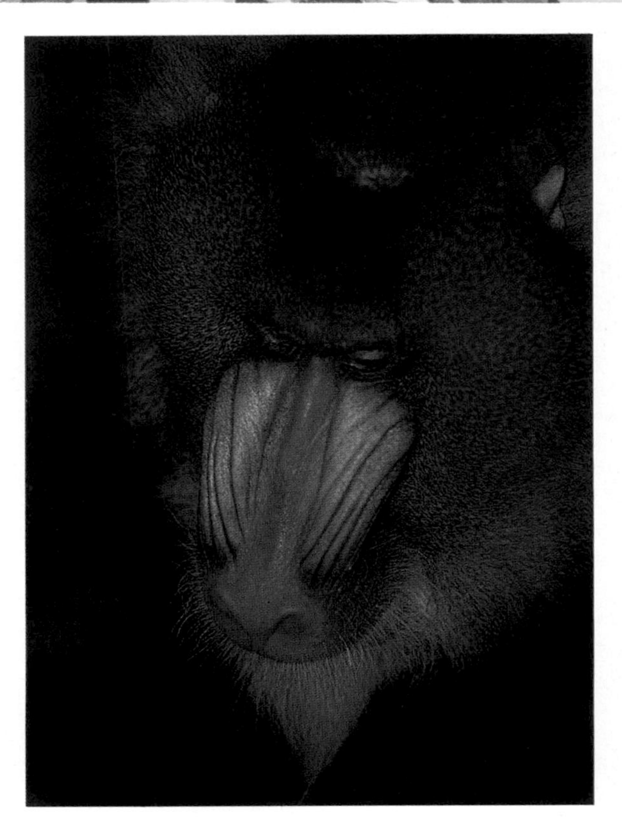

South America

Top: one of the strangest and most feared animals, the vampire is the only blood-sucking mammal. It lands on or near its victim, often a horse or cow but sometimes a sleeping human. Then it slices out a chunk of flesh with its razor-sharp teeth, pours anticoagulant saliva into the wound and laps up the blood.

Above: using its strong claws, a giant anteater tears open a termite nest and inserts its long sticky tongue.

In much the same way as the mammals of Australia are descendants of survivors that reached the continent before it separated from the rest of the world (page 20), so many of the mammals of South America evolved in isolation from North America. About 60 million years ago, South America became cut off as an island. The mammals already there included various marsupials, the armadillos, the sloths, as well as many types that are now extinct. Bats joined them by flying across the gap from North America, and rats and monkeys by floating. When the land link through Central America rose again from the sea 50 million years later, deer, peccaries (relatives of the pigs), vicunas (relatives of the camels), carnivores, horses and tapirs invaded the south, while opossums, armadillos and ground sloths spread northwards. Since then the horses, ground sloths and some other types have become extinct in America, although horses were later re-introduced.

Among the unique mammals of South America, there is the order of the edentates, the toothless mammals, a name that is rather misleading because some do possess teeth. Apart from armadillos, which have spread into North America, all modern edentates live in South America. At one time there was a much larger selection including the giant ground sloths and the glyptodonts, which looked rather like enormous tortoises. Many edentates have since become extinct, leaving only the anteaters, tree sloths and armadillos.

The giant anteater is about three metres long, of which one metre is a bushy tail. It lives on the ground in bush country, whereas the tree anteaters, or tamanduas, and the squirrel-sized pygmy anteater are tree dwellers. All anteaters feed mainly on termites and ants. They rip open the nests with strong claws, thrust in their long snouts and lick up the insects with their sticky tongues. The same specialisations for eating ants and termites have been evolved by other mammals: the pangolins of Africa and Asia, the aardvark of Africa and the echidnas of Australia.

The sloths spend their lives hanging upside down from branches. Their fur lies from belly to back, so that rain water still runs off their coats; their claws are hook-shaped. As their name suggests, they are very slow-moving and spend 18 hours a day asleep. Armadillos are the only living armoured mammal. The armour is made of plates and bands and is flexible enough for some species to be able to roll into a ball.

The camel family evolved in North America. One branch migrated across the Bering Straits to the Old World and is represented by the Arabian and Bactrian (two-humped) camels. A second branch moved into South America to become the wild guanaco and vicuna and domesticated llama and alpaca. All camels live in dry country, and the South American camels live in the high Andes.

South America is the home of many rodents not found elsewhere. The most familiar are the guinea pigs, first domesticated by the Incas for food. Their wild relatives are the cavies and others include the agouti of forested regions, the viscachas and hare-like maras of the plains and the tuco-tucos ranging from the plains to the mountains. The chinchillas, a squirrel-like animal of the Andes, and the beaver-like coypus of rivers and swamps are farmed for their fur. The capybaras are the largest living rodents. Pig-sized, they live in groups near water and are good swimmers.

Opposite bottom: the puma, sometimes called cougar or the mountain lion, ranges throughout America, from Canada to Patagonia. It lives in a variety of habitats and preys on deer and other animals.

Top: the three-toed sloth feeds on leaves and fruit. It often has a greenish appearance caused by algae growing on its hair, which is also the home for a certain moth.

Above: the nine-banded armadillo has spread into North America during the last century. These animals are unique among mammals because the female always bears identical quadruplets.

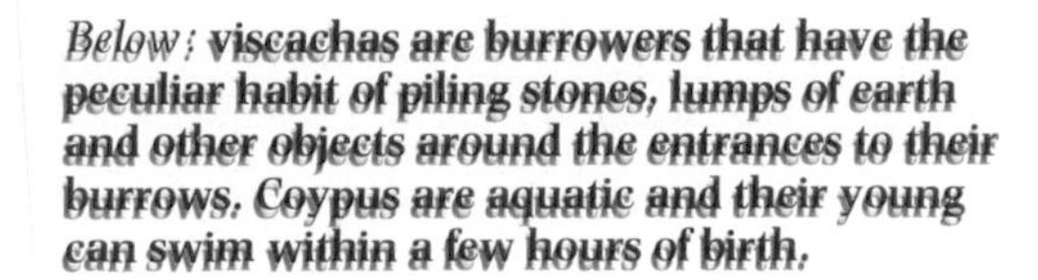

Below: viscachas are burrowers that have the peculiar habit of piling stones, lumps of earth and other objects around the entrances to their burrows. Coypus are aquatic and their young can swim within a few hours of birth.

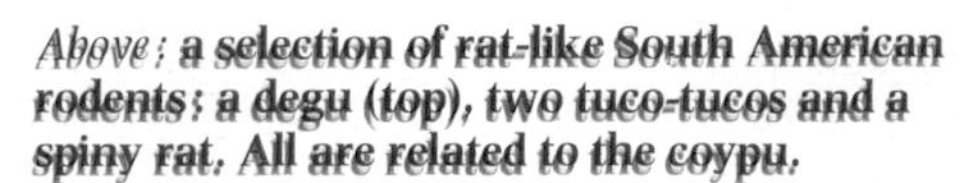

Above: a selection of rat-like South American rodents: a degu (top), two tuco-tucos and a spiny rat. All are related to the coypu.

Whales and dolphins

The whales and dolphins are often collectively called the cetaceans, as there is no common English name for the order Cetacea. The name whale is given to the larger species, even though the killer whale is a member of the dolphin family. Porpoises are members of the porpoise family Phocaenidae, but in North America the name is given to some dolphins.

The cetaceans are the most aquatic of all mammals. They are so well-adapted for life in water that they die if they become stranded. The body is very fish-like. Resistance to the water flow is reduced by the lack of hair, external ears and other obstacles, and it is also given a streamlined shape by the layer of blubber, which acts as an energy store and keeps the animal warm. The forelimbs have become flippers for steering, the hindlimbs have disappeared, and the tail has broad flukes, which beat up and down to propel the animal.

Unlike fishes, which breathe underwater by means of gills, cetaceans show their descent from land mammals by breathing air through lungs. They must come to the surface to breathe through their blowholes, which are nostrils situated on the top of the head. Some cetaceans can descend to great depths, and the record is held by a sperm whale which was found entangled in a telegraph cable at 1135 metres. Whales do not suffer from bends like human divers, because they take only one breath of air down in their lungs, so there is very little nitrogen to form dangerous bubbles in the bloodstream. Cetacean calves are born tail first and have to rise to the surface immediately to take a breath and fill the lungs for buoyancy.

There are two main kinds of cetaceans: the baleen whales, Mysticeti, and toothed whales, Odontoceti. The baleen whales have rows of bristly plates called baleen, or whalebone, hanging on each side of the mouth from the gums of the upper jaws. The plates are used to strain swarms of small animals, ranging from minute shrimp-like crustaceans to fish and squid, from the sea. The right whales have baleen plates up to three metres long, and they swim with their mouths open and water pouring between the plates. Other baleen whales, like the blue whale and the humpback whale, take huge gulps of water then shut their mouths and squeeze the water through the baleen. In both cases the baleen strains out the mass of small animals which is then swallowed.

The toothed whales are more numerous than baleen whales. There are 66 species, compared with 10, and they include the large sperm whale, the many dolphins and porpoises, the killer whale and the pilot whale. Their diet is mainly fishes and squid, but the killer whale also eats seals, dolphins and seabirds. The killer whale has a full set of sharp, triangular teeth, but the sperm whale has teeth only in the lower jaw and some species have no more than two teeth. The narwhal has a single outsize tooth – the tusk – which is growing through the upper lip.

Right: **the adult male killer whale is recognised by its tall dorsal fin as well as by the colour pattern of black and white. There are usually one or two adult males in each group, or pod.**

Opposite top: a grey whale rears out of the water to survey the scenery.

Opposite centre: the humpback whale has many similarities to the rorqual. The picture shows the whale's characteristic long pectoral fin. Despite its size, the humpback may often be seen leaping and rolling back into the sea.

Right: the sperm whale is the largest of the toothed whales and is a deep water species; it can dive up to 450 metres in search of its favourite food, the giant squid.

Below left: the common dolphin (centre), the narwhal (left) and the porpoise (right).

Below right: representatives of the three families of baleen whales: grey whale, lesser rorqual or minke whale, and Greenland right whale. These whales depend on curtains of baleen plates for sieving small animals from the water. Rorquals take huge gulps of water and then squeeze it through the baleen.

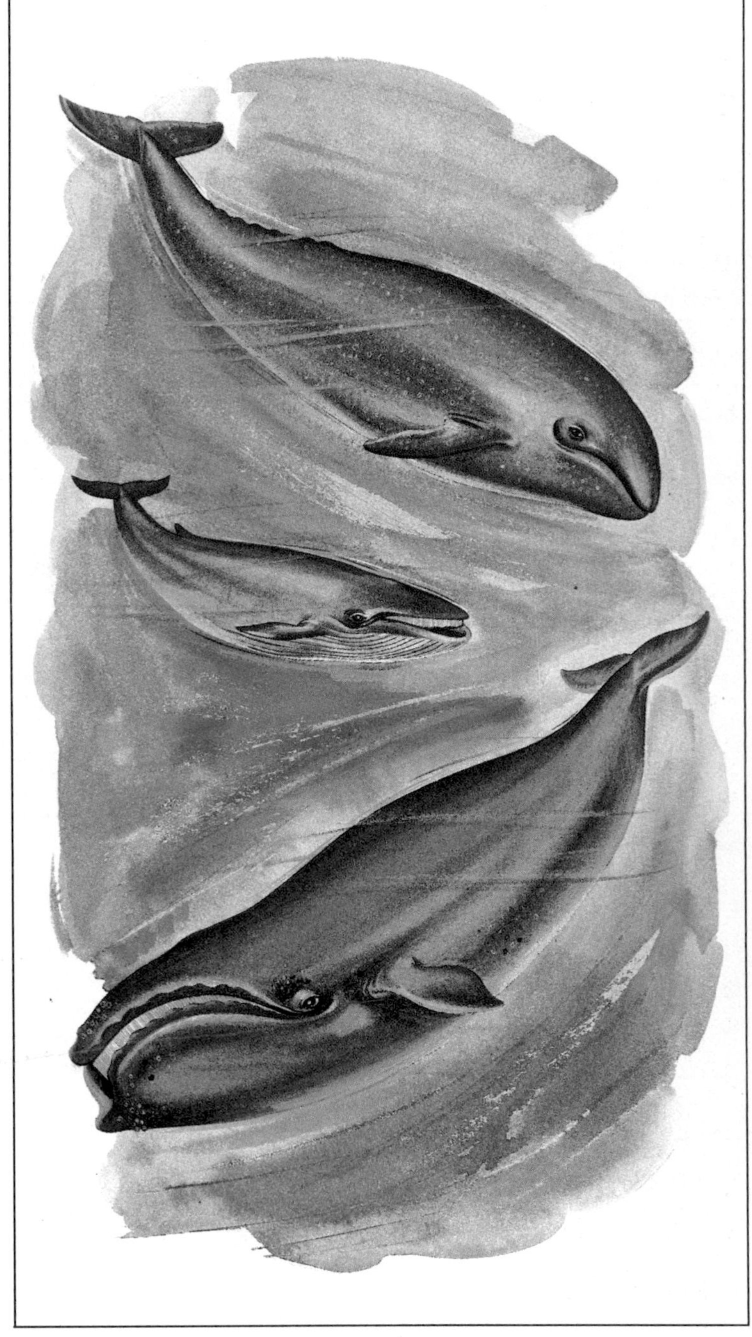

Seals

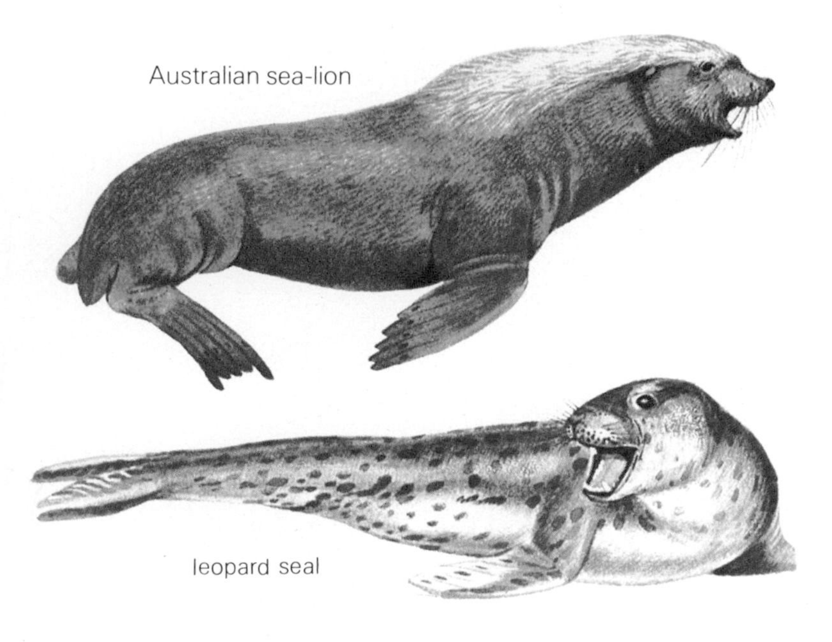

Australian sea-lion

leopard seal

sea-lion

common seal

Australian fur seal

Seals are streamlined, agile swimmers which come out of the water only to breed and bask. They can dive to great depths in search of food, and some are known to use echolocation for navigating in dark waters. They are basically fish eaters, but they eat other kinds of seafood. Elephant seals *(right)*, for instance, take large numbers of squid, and crabeaters *(centre right)* eat shrimp-like krill. Leopard seals often hunt penguins.

Seals are carnivores that have returned to the sea. They have streamlined bodies, and the limbs have become flippers for swimming. There are two main kinds of seals. The true, or hair, seals are descended from an otter-like ancestor. They swim by side-to-side movements of the hind-flippers, and they crawl clumsily on land. The eared seals, which include the sea-lions and fur seals, and the walrus are descended from a bear-like ancestor. They swim with up-and-down movements of the fore-flippers. The eared seals can run as fast as a man by turning their hind-flippers forward and lifting the body off the ground.

Seals eat mainly fish, but some species also eat crustaceans and squid. The crabeater seal of the Antarctic eats the crustacean called krill, which it strains from the water by using its teeth as a filter. The leopard seal, also of the Antarctic, eats small seals and penguins as well as fish.

Some seals are known to search for food at great depths. The Weddell seal regularly dives to 100-200 metres and the deepest dive on record is 600 metres. One Weddel seal stayed under for 70 minutes. Other seals feed mainly at night when fish, squid and other animals come nearer the surface. Seals can dive for a long time because their muscles can work anaerobically (without oxygen) for longer than those of land mammals. Also, they conserve oxygen by reducing the heart beat from 50-100 per minute to 10 or less, and blood is directed mainly to vital organs, such as the brain, by shutting off flow to the disgestive organs and muscles.

Once a year seals gather at traditional breeding grounds to bear their pups. Each cow bears one pup and mates again within a few days. In the eared seals, walrus and a few true seals, such as the elephant seals and grey seal, the bulls divide the breeding area into territories. Only the largest bulls can get a territory, which they defend against other bulls so that they can mate with the cows that have given birth within its boundaries. In these species the bull is much larger than the cow: an elephant seal bull weighs up to 3600kg and a cow only 900kg.

The cows of true seals suckle their pups for a short time before abandoning them. The harp seal pup is suckled for only nine days. During this period, the cow does not feed and she rapidly loses weight from the manufacture of milk which is 50 per cent fat (compared with five per cent in cow's milk). Eared seals and the walrus rear their pups more slowly; weaning in some species may take one year or more, and the cows go to sea to feed at intervals.

Above: **the most abundant species of eared seals is the California sea-lion.**

Centre: **fur seals make up the second group of eared seals and they have a thick underfur.**

Right: **walruses use their tusks for fighting, digging up molluscs and for hauling themselves out onto ice flows.**

Domestic mammals

Above: **goats were first domesticated at least 9000 years ago, and they are still important in dry parts of the world. Unfortunately, through their ability to eat the toughest scraps of vegetation they have aided the spread of deserts. They may also cause damage in fruit orchards.**

Below: **the domestic pig is a prolific source of meat. Unlike other domestic animals, pigs cannot be herded. They are either penned in sties or allowed to run free.**

The origins of domestic animals will never be known because they stretch back to the Stone Age when there were no written records. The remains of bones do not help because of the difficulty of distinguishing between wild and domestic animals. It is often not even possible to separate sheep and goats, and it is possible that sheep are descended from goats.

Humans made the transition from hunters to farmers when they realised that it was easier to rear animals than to hunt them. Domestication was probably started by bringing home orphaned animals and keeping them until they were full grown. The animals acted as living food stores. Some mountain people still feed their surplus grain to goats which are eaten later. Another possible course of domestication was through the habit of nomadic hunters to follow herds of animals, as the Lapps follow their reindeer, and protect them from predators in return for killing a few for meat and skins. Gradually the men and the herds would become more dependent on each other.

Domestic cattle are descended from a wild ancestor, the aurochs. This became a crop robber and was brought into domestication instead of killed. The wolf, the ancestor of the dog, was similarly attracted to human settlements in search of scraps. It then accompanied hunting parties and gave warning of intruders.

Once mammals started to breed in captivity, their keepers could select certain traits. Cattle, for instance, were produced with smaller horns and shorter stature so that they were less dangerous. By 6000 years ago the civilisations of the eastern Mediterranean had produced distinct breeds of cattle.

Out of the thousands of species of mammals only a handful have been domesticated; the commonest are cattle, horses, donkeys, sheep, goats, pigs, cats and dogs. As beasts of burden in different environments, camels,

water buffaloes, llamas, yaks, elephants, reindeer and even dogs and sheep have been pressed into service. Rabbits and guinea pigs have been bred for the table, and there have been numerous attempts at domestication which have not lasted. The ancient Egyptians fattened hyaenas and gazelles, and the Romans kept dormice. Mongooses and genets have been kept like cats to keep down vermin. Modern attempts at domestication have involved red deer and eland, as well as rats and monkeys for medical research, and mink and coypu for fur.

The key to successful domestication is that the animal should be easy to manage and breed. Successful farm animals are those which naturally live in herds that accept the domination of a leader and a life of daily routine.

Bottom centre: **banteng (bull and cow on left) and gaur (top right) are wild species of cattle. The water buffalo (bottom right) is domesticated, very few being left in the wild.**

Below: **zebu cattle are descended from Indian wild cattle. They have a hump on the shoulders.**

Bottom right: **husky-drawn sledges are a favourite form of polar transport, but dogs have also been used for pulling carts or carrying loads on their backs.**

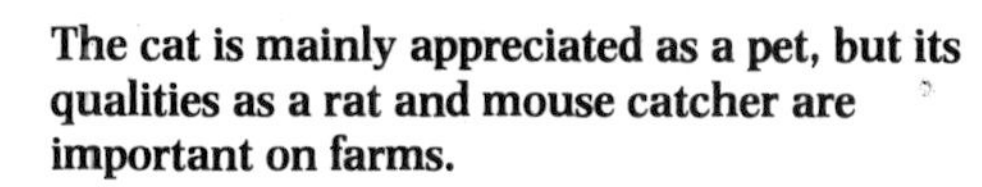

The cat is mainly appreciated as a pet, but its qualities as a rat and mouse catcher are important on farms.

Centre right: **the horse was once an essential animal for farm work and transport.**

Mammal pests

Above: the coypu was brought to Europe and bred on ranches for its fur. Many coypus have escaped and become pests by burrowing through river banks and eating crops.
Left: unlike other bats, the vampire is a serious pest. By sucking the blood of cattle it can seriously weaken them, and it may also transmit rabies.
Below: goats feed on the sparsest vegetation and can overgraze an area– a serious problem where they have been freed on tropical islands. They may also become scavengers in towns.

As human beings started to exploit the world by agriculture, forestry and fishing, so they came into conflict with mammals. Sometimes the problem is local, as with humpback whales getting tangled in nets off Newfoundland, palm civets stealing palm juice being collected to make toddy, polar bears harassing Arctic townships, and bats fouling altar cloths with their droppings.

Elsewhere, the damage is widespread and serious. Gerbils can ruin a maize crop by removing the seeds as they are planted; rabbits and goats destroy vegetation so effectively that they bring complete desolation. In these and many other instances, such as seals stealing salmon, wolverines attacking reindeer herds and hedgehogs stealing partridge eggs, the mammals are pests because they compete for things that humans would like to keep for themselves.

The scale of the damage can be enormous. The worst culprits are the rodents, which eat or spoil both growing and stored crops and then attack prepared foodstuffs. So far there is no satisfactory method of getting rid of rodents; with a high birth rate and the ability to migrate, replacements soon recolonise cleared areas. On the other hand, it is a matter of regret for many people that many large mammals are wiped out all too easily. The wild ancestors of horses, cattle and goats, which compete with the domestic forms for food and space, are now either extinct or very rare.

Mammal pests can transmit a number of diseases to humans and domestic animals. Plague, the Black Death, is carried by rats and transmitted by their fleas, and the vervet monkey is the carrier of the recently discovered, rare and fatal 'green monkey disease'. Rabies is a particularly unpleasant disease that is usually associated with the bite of a 'mad dog' in which the disease is already well advanced, but it is also carried by wild animals, notably foxes, in some parts of the world.

The damage caused by mammal pests is astronomical and huge sums are spent on control. Where a mammal pest is a serious problem, there is little chance of it being totally destroyed or even reduced to a level where it causes insignificant damage. The only solution is to reduce its numbers as much as possible. In some cases, as with seals eating fish, it is not always certain that removing the pest will increase the crop available to humans. An essential part of modern pest control is to study the pest's habits either to see how it may be attacked most efficiently or to reduce its depredations. For instance, shutting up chickens securely at night is a better plan than trying to kill all the neighbourhood's foxes.

The grey squirrel was introduced to Britain because it is an attractive mammal, but it became a nuisance by damaging trees.

Below left: **the brown rat does enormous damage around the world by eating crops and stored food. It also carries disease.**

Below: **the red fox has always been a pest to farmers and gamekeepers, but it is now a more serious pest because it spreads rabies.**

The future of mammals

As the population of one mammalian species, man, continues to grow at an alarming pace, so pressure is put on all other mammals, apart from a few that benefit directly from human activities.

The wild places of the world are gradually shrinking, so there are fewer habitats for mammals. This is most obvious in the tropical rain forests, which are disappearing rapidly. Their animal inhabitants cannot survive without the trees that provide them with food and shelter. In Madagascar there is very little left of the forest which once covered the island. Where there are remnants the native lemurs, carnivores and tenrecs survive; where there is now open country there are no mammals. It is believed that many species of bats in rain forests are becoming extinct before scientists have even had time to name them, and only a few types of mammals, such as some rodents and monkeys, can adapt successfully to the plantations of palms or rubber trees which replace the forests. In other areas the inhabitants have a better chance of survival because the changes are not so great, but there will be a reduction in the variety of mammals and most species will survive in fewer numbers.

The second threat to mammals is interference by man. This includes the slaughter of animals, like the chasing of the Arabian oryx in motor vehicles for sport, or the protection of farm animals and crops from wild mammals, and the killing of animals for particular products, such as elephants for ivory or rhinoceroses for horns. The quagga, a once common kind of zebra, became extinct in 1878 and the thylacine, a marsupial carnivore, has not been seen for years.

Increasingly, mammals are being forced into retreat and their only sanctuary is in reserves and parks. Although we still think of the African savannahs as being full of elephants, lions, giraffes, zebras and many antelopes, in reality they are surviving only

Above: **the history of the Arabian oryx is a success story for conservation. When hunters had killed the last wild oryx, there were enough in captivity to breed sufficient numbers for some to be released into their original habitats.**

Left: **the giant panda is a symbol of conservation. However, although it is protected by the Chinese people, it may be seriously affected if its bamboo food supply is lost.**

Opposite centre: **gorillas are threatened in their reserves by poachers who sell their skulls and hands as souvenirs.**

Opposite bottom left: **despite international outcry, the hunting of whales continues and the numbers of some species remains low.**

Opposite bottom right: **vicunas of the Andes are now ranched so that their numbers can increase while local people also benefit.**

Above: **hunting and the destruction of the habitat made tigers very rare in India, but the setting up of huge reserves for them have reversed this trend.**

where they are protected. So much of the rest of Africa is under cultivation that there is no room for large animals. Where this is the situation the animals are virtually living in a huge zoo. They are not leading a free, natural life but only surviving by courtesy of man. They have to be protected from poachers and, if necessary, culled when they become too numerous in the park.

Nevertheless, parks and wildlife management have saved several mammals. The bison very nearly disappeared from the American prairies but it is now flourishing in reserves. The Arabian oryx became extinct in the wild but was saved in zoos, and captive bred oryxes have been released in their native homes again. Reserves now provide safe homes for many other kinds of animals.

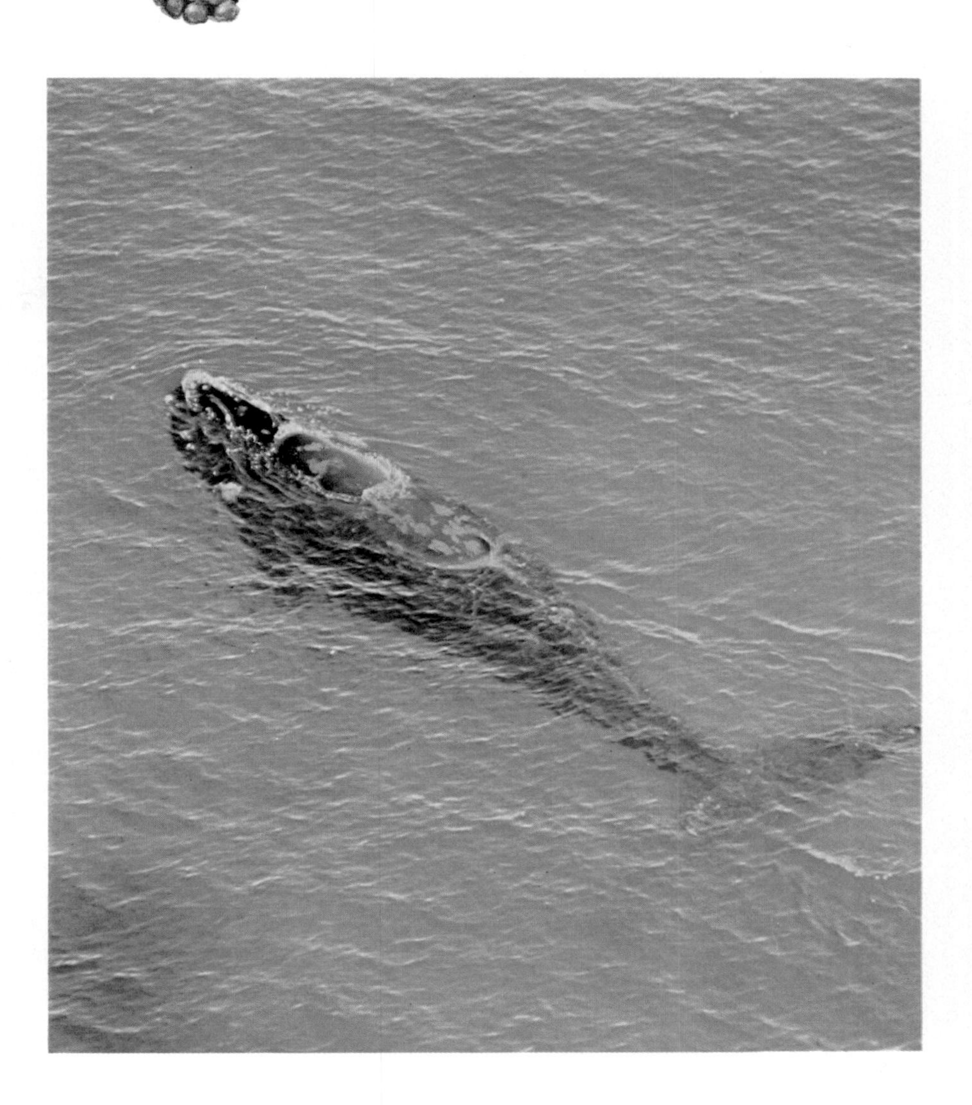

Mammal-watching

Mammals are shy creatures and many come out only at night, so mammal-watching is more difficult than bird-watching. It is easy to go out with binoculars and find some birds to study, even if they are only common species, but hours of watching and waiting for mammals may lead to nothing but disappointment.

Professional mammalogists used to see mammals only when they caught them in traps, but the uses of night-viewing devices and of miniature radio transmitters attached to animals to track their movements have revolutionised the study of mammal habits. However, these are expensive, specialist devices and the amateur mammal-watcher must still use traditional methods.

The only mammals that can be watched satisfactorily are those which come out by day and have regular habits. Squirrels, hares, rabbits and deer come into this category, and observations can be made once their regular haunts are discovered. Water voles are easy to see if their waterside burrows and grazing patches are found, while watching badgers at their setts is one of the most popular forms of mammal-watching. Putting out food will attract mice and hedgehogs into view.

In general, however, more can be learned from observing signs than seeing the mammals themselves. Tracks can be followed and much can be learned from prey remains and droppings, to piece together a picture of the mammal's way of life.

Because badgers are large and conspicuous, live in traditional burrows, or setts, and emerge regularly around sunset, they are quite easy to observe. However, like most mammals, they are shy and it is easier to learn their habits by searching for signs of their activity.

Index

Page numbers in *italics* refer to illustrations